BIODIVERSITY

Edited by **Adriano Sofo**

Biodiversity
Edited by Adriano Sofo

Published by InTech
Janeza Trdine 9, 51000 Rijeka, Croatia

Publishing Process Manager Dragana Manestar

Technical Editor Teodora Smiljanic

Cover Designer InTech Design Team

Additional hard copies can be obtained from orders@intechopen.com

Biodiversity, Edited by Adriano Sofo
p. cm.
ISBN 978-953-307-715-4

For EU product safety concerns:
IN TECH d.o.o., Prolaz Marije Krucifikse Kozulić 3, 51000 Rijeka, Croatia,
info@intechopen.com or visit our website at intechopen.com.

Meet the editor

Dr. Adriano Sofo, PhD, is working as an Assistant Professor at the University of Basilicata (Italy). He has experience in teaching and research at the Universities of Bari, Potenza and Bologna (Italy), Hasselt (Belgium), Uppsala (Sweden), Toulouse (France); in other institutes of public and private research, such as ENEA (Italy), Geisenheim Research Center (Germany), and other Agricultural Experimental Institutes. His research fields are plant and soil biochemistry and molecular biology, plant eco-physiology, and plant cell and tissue cultures, with a particular emphasis on sustainable soil management and plant responses to drought as well as to other abiotic and biotic stresses. Dr. Sofo is a lecturer in soil science and plant physiology at University of Basilicata and he is the author of approximately 50 papers published in international books and journals. He is a referee for a series of peer-reviewed international scientific journals in his research sectors.

Contents

Preface

Biodiversity is strongly affected by the rapid and accelerating changes in the global climate, which largely stem from human activity. Anthropogenic activities are causing highly influential impacts on species persistence. The sustained environmental change that wildlife is experiencing may surpass the capacity of developmental, genetic, and demographic mechanisms that species have developed to deal with these alterations. How biodiversity is perceived and maintained affects ecosystem functioning, as well as the fact how the goods and services that ecosystems provide to humans can be used. Recognizing biodiversity is essential to preserve wildlife. Furthermore, the measure, management and protection of ecosystem biodiversity requires different and innovative approaches.

This book is divided in three sections. The first two correspond to the different levels at which biodiversity can be measured: ecosystems or organisms. The knowledge of species distribution is a vital component in wildlife conservation and management. Such information aids in quantifying organism–habitat relationships, describing and predicting differential space use by animals, and ultimately identifying habitat that is important to an organism. A study of this has produced a variety of models that combine observations of species occurrence or abundance with environmental estimates, based on statistically or theoretically derived information (Chapter 1). Internal and external factors of change seem to be currently degrading and homogenizing the biodiversity of many ecosystems, as in the case of Mediterranean cultural landscapes. Indeed, many results show decreased capacity of Mediterranean ecosystems to provide regulation services, a process that has continued in spite of the conservationist policies implemented during several decades (Chapter 2). At organism level, invertebrate diversity seems to be strongly affected by the amount of biomass, and in particular by deadwood. For this reason, it is necessary and important to determine the positive effects of deadwood on invertebrate diversity (Chapter 3). A series of molecular techniques, such as flow cytometry and biopanning, were recently discovered and used for *in vitro* studies of proteins on the surface of bacteria. All these tools are of key importance for estimation of bacterial diversity, that plays a key role in affecting biodiversity at the higher levels of terrestrial and aquatic trophic chains (Chapter 4). Regarding the latter, the reduction of Cr(VI)-Cr(III) in the environment is beneficial to ecosystems since Cr(VI) is highly toxic and mobile in aquatic systems but, in certain groups of bacteria, the Cr(VI) reduction capability may be transferred across different species. Successful simultaneous removal of Cr(VI) with organic co-pollutants demonstrated the potential of biologically engineering microbial species to clean up environments contaminated with a range of diverse pollutants, so preserving ecosystem biodiversity. Therefore, Chapter 5 evaluates the prospects of application of the biological remediation against Cr(VI) pollution and recent improvements on this fundamental process.

The last section of this book is focused on the molecular techniques used for measuring biodiversity, a critical point of the studies on biodiversity. Indeed, with molecular and analytical techniques (FISH, DNA-microarray, etc.) now we can begin to understand how marine biodiversity supports ecosystem structure, dynamics and resilience. With these innovative techniques, it is possible to augment the understanding of biodiversity and ecosystem dynamics in all areas of the planktonic community. The authors of Chapter 6, review selected molecular techniques and provide case studies to illustrate their use for biodiversity purposes. One of the possibilities to measure biodiversity is to use DNA, as it is universal, relatively stable, suitable and reliable for measures, and comparable among a broad range of organisms. The increasing amount of data deriving from DNA sequencing it is not easy to manage, and the choice of good molecular markers should consider the species to be studied for specific biodiversity analysis.

The aim of the present book is to give an up-to-date overview of the studies on biodiversity at all levels, in order to better understand the dynamics and the mechanisms at the basis of the richness of life forms both in terrestrial (including agro-ecosystems) and marine environments. On this basis, the present volume would definitely be an ideal source of scientific information to the advanced students, junior researchers, faculty and scientists involved in ecology, agriculture, plant and animal sciences, environmental microbiology, molecular biology, biochemistry, biotechnology and other areas involving biodiversity studies.

I am thankful to all the contributors for their interests, significant contributions and cooperation that made the present volume possible. I also thank Prof. Antonio Scopa and Prof. Cristos Xiloyannis. Without their unending support, motivation and encouragements during all my years of academic career the present grueling task would never have been accomplished.

Adriano Sofo, PhD
University of Basilicata Potenza
Italy

Part 1

Ecosystem-Level Biodiversity

1

Integrating Spatial Behavioral Ecology in Agent-Based Models for Species Conservation

Christina A.D. Semeniuk, Marco Musiani and Danielle J. Marceau
University of Calgary
Canada

1. Introduction

Anthropogenic activities are causing highly influential impacts on species persistence. The sustained environmental change wildlife are experiencing may surpass the capacity of developmental, genetic, and demographic mechanisms that populations have evolved to deal with these alterations. Undeniably, habitat fragmentation, habitat loss, and human disturbance are causing a decline in species numbers on a global scale, with shifts or reductions occurring in species-distribution ranges. The knowledge of species distribution is a vital component in wildlife conservation and management. Such information aids in quantifying animal–habitat relationships, describing and predicting differential space use by animals, and ultimately identifying habitat that is important to an animal (Beyer et al. 2010). The field of species distribution modeling (SDM) as a means of quantifying species–environment relationships has been extensively developed since the first formal definition of differential habitat selection theory by Fretwell and Lucas in 1969. It has since produced a variety of numerical tools that combine observations of species occurrence or abundance with environmental estimates based on statistically or theoretically derived response surfaces (Guisan and Zimmermann 2000). These models include presence/absence models, dispersal/migration models, disturbance models, and abundance models; they are now widely used across terrestrial, freshwater, and marine realms.

SDMs are used to determine the suitability of the organisms' habitat, relying on density/abundance measures or the ratio between used and available habitats to infer habitat quality. These models use spatial environmental data to make inferences on species' range limits (Kearney and Porter 2009). Most approaches are correlative in that they statistically link spatial data (typically geographic information systems data) to species distribution records. Despite the prevalence of SDMs in applied ecology, a review of recent papers cautions using a statistical description that implicitly captures these "habitat use" processes as they are statistically associated with the predictor variables, but may not be so biologically. Firstly, habitat use does not necessarily equate with high quality habitat, range requirements, nor resultant increased wildlife fitness because biotic and abiotic cues can cause animals to choose habitats that do not provide the necessary resources to ensure high fitness returns (Jonzén 2008; Pérot and Villard 2009). Secondly, SDMs are frequently applied for predicting potential future distributions of range-shifting species, despite these models' assumptions that (1) species are at equilibrium with the environments, and (2) the data used to train (fit) the models are representative of conditions to which the models are already statistically associated, and not to which they are anticipated (Elith et al. 2010). Animal responses to novel environments, therefore, especially ones that may be a mismatch to the habitats in which the animal evolved, can render the predictions of SDMs ineffectual. A lack of insight into the processes that govern animal movement and habitat selection can have consequences on the predictive success of SDMs in determining range limits and habitat suitability. This can then have carryover effects on the resolving of spatial issues (such as extent and resolution, geographical- and environmental space), and the statistical methodologies used to test model fit and selection.

Methodological innovations have been recently proposed to improve the predictability of conventional SDMs in spatial modeling of animal-habitat interactions. These newer models incorporate explicit relationships between environmental conditions and organismal performance, which are estimated independently of current distributions. They include: (i) the integration of animal movement and resource-selection models to arrive at biologically-based definitions of available habitat (Fieberg et al. 2010), (ii) the use of state-space movement models (Patterson et al. 2008), (iii) linking species with their environment via mechanistic niche modeling (Kearney and Porter 2009), (iv) and combining resource-selection functions, residency-time and interpatch-movement analyses (Bastille-Rousseau et al. 2010). These emergent efforts have one common, unifying feature: the need to implicitly or explicitly incorporate mechanism; that is, the underlying physiological, behavioral, and evolutionary basis for animal movement and habitat use. The emphasis on improving the statistical fit of SDMs via the incorporation of more ecologically-relevant procedures highlights the multiple advantages when considering the mechanistic links between the functional traits of the organism and its environment. These are: (1) the understanding of the proximate constraints limiting distribution and abundance, (2) the examination of the ultimate consequences of species range effects and population persistence, and (3) the exploration of how organisms might respond to environmental change.

One of the challenges in incorporating mechanism into SDMs is that these models can be limited by the availability of data for model parameterization and because their success in predicting range limits relies on the identification of key, abiotic limiting processes, such as climatic factors, humidity, etc., that have both proximate and ultimate effects on species distributions (Elith et al. 2010). These limiting processes, or constraints, might not be the most important ones, or equally important, in all areas of a species' range. In addition, the interaction between different abiotic constraints and those between abiotic and biotic constraints could cause observed ranges to deviate from predicted ranges. In essence, emergent relationships between the organism and a changing environment cannot be captured by mechanistic SDMs. Lastly, few studies have explicitly incorporated geographic variation in animal traits or genetic variation across a range in mechanistic models, thus essentially ignoring that unique phenotypes may behave in significantly different ways. For a more comprehensive review of correlative versus mechanistic SDMs, we refer the reader to Buckley et al. (2010). In this paper, we present an alternative approach to conventional correlative and mechanistic species distribution modeling, called agent-based modeling that can be used as an effective tool for understanding and forecasting animal habitat selection and use. This methodology offers several advantages. First, it can accommodate ecological and evolutionary theory in the form of behavioral ecology. Second, it can be readily integrated with the concepts of spatial ecology. In doing so, agent-based models (ABMs) can redress the fundamental issues of mechanism, spatial representation, and statistical model evaluation. ABMs can thus enable the exploration of how wildlife might respond to future changes in environmental conditions - an inquiry of utmost importance for wildlife conservation and management.

This chapter is organized as follows. First, we begin by discussing why animal behavior should be incorporated into studies of wildlife conservation, and how its oversight can lead to erroneous understandings and predictions of critical habitat. We then describe how behavioral ecology provides the basic understanding of the mechanisms driving animal habitat selection and dispersal/migration behaviors; and we argue that it should be incorporated with the concepts of spatial ecology and its geospatial tools. Next, we introduce agent-based modelling and demonstrate how it represents the ideal framework for assimilating behavioral mechanisms with temporal-spatial processes to drive animal movement and habitat selection, and to determine habitat suitability and species distribution. Based on this principle, we then show how the incorporation of spatial behavioral ecology in ABMs can address issues of scale commonly found with the more conventional species-distribution models with regards to extent and resolution, and geographical and environmental space. We also discuss the issues of statistic evaluation of best fit models. We conclude by summarizing the potential of ABMs for wildlife conservation planning, and by suggesting areas for improving their flexibility and performance.

2. Behavior as a key mechanism

2.1 The advantages of addressing behavioral mechanisms over choosing statistical empiricism

As mentioned above, statistical statistical SDMs perform poorly in identifying true habitat quality when the mechanisms driving habitat selection are not explicitly incorporated into the modelling process. This is because strong social interactions, temporally unpredictable habitats, post-disturbance crowding effects, non-ideal habitat selection, and ecological traps all lead to animals either under- or over-utilizing a habitat that produces greater or fewer fitness returns than others available on the landscape, respectively (Johnson 2007; Jonzén 2008). For instance, Mosser et al. (2009) found that density was a misleading indicator of lion (*Panthera leo*) habitat quality in the Serengeti, as this metric identified 'source', high-quality sites that were actually low-quality sites that merely provide refuges for non-reproductive individuals. Over a multi-year and multi-site study of yellow warbler (*Dendroica petechia*) nest microhabitat selection, Latif et al. (2011) found a consistently negative relationship between preferred microhabitat patches and nest survival rates, suggesting that maladaptive nest microhabitat preferences arose during within-territory nest site selection. The authors attribute this mismatch to the recent proliferation of the parasitic brown-headed cowbird (*Molothrus ater*), and/or anthropogenic changes to riparian vegetation structure as likely explanations. These behavioral phenomena will result in SDMs identifying habitats as being suitable foraging, breeding, or dispersing grounds, when in fact there has been a mismatch between habitat use and fitness, with serious ramifications for conservation planning.

Novel or disrupted environments can also violate the assumption of correlational SDMs that animal populations are at equilibrium. Ecological niches may expand or go extinct, affecting population demographics and species ranges via animal behavior in discontinuous or non-linear ways. Schtickzelle et al. (2006) studied how habitat fragmentation modified dispersal at the landscape scale in the specialist butterfly *Proclossiana eunomia*. They showed that dispersal propensity from habitat patches and mortality during dispersal were the consequences of two different evolutionary responses of dispersal behavior. They concluded that evolutionary responses can generate complex nonlinear patterns of dispersal changes at the metapopulation level according to habitat fragmentation, making predictions of metapopulation effects challenging. Additionally, the success or failure of establishing populations, or altering animal distributions in different environments is mediated by animals that benefit from the presence of conspecifics or heterospecifics after settlement, or are governed by personality-dependent dispersal. In a long-term study of the range expansion of passerine birds, Duckworth and Badyaev (2007) concluded that the coupling of aggression and dispersal strongly facilitated the range expansion of western bluebirds (*Sialia mexicana*) across the northwestern United States over the last 30 years. As such, forecasting the responses of wildlife to changes in their environment without acknowledging the mechanisms involved can give potentially misleading predictions of range effects.

2.2 Conservation behavior as a discipline

Conservation behavior is a relatively new interdisciplinary field aimed at investigating how proximate and ultimate aspects of animal behavior can be of value in preventing the loss of biodiversity (Bushholtz 2007). Animal behavior is an important determinant in species persistence since how an animal behaves determines its survival and reproductive success. In particular, natural selection favors individuals who adopt life history strategies that maximize their gene contribution to future generations. Expression of these strategies typically manifests itself through the behaviors of the animal that possess a heritable component sufficient to allow natural selection to operate. Thus, the behaviors of animals attempting to maximize their lifetime fitness will affect survival, reproduction, and hence recruitment, ultimately scaling up to the population level and species persistence.

Indeed, many of the initial responses by animals to environmental change are behavioral i.e., changes in feeding location, prey selection, or movement responses to disturbance. Behavioral indicators can provide an early warning to population decline or habitat degradation before numerical responses are evident. Similarly, they can be used to monitor the effectiveness of management programs, or evaluate the success of a management program at its early stages, before population or ecosystem-level responses are evident (Berger-Tal et al. 2011). While these concepts may seem atheoretical and merely descriptive, there is a strong incentive to understand the underlying motivations involved in animal responses to anthropogenic impacts and their mitigation. As an illustrative example, when managers plan for critical habitat, it is imperative to ensure: (1) that enough cover is present so that the animal does not spend an excessive amount of time being vigilant at the expense of acquiring its energetic requirements, (2) that the food resources available will not cause the animal to spend excess time searching or assimilating their forage at the expense of other activities such as dispersing successfully, breeding or caring for young, (3) that animals are not crowded into habitats so that foraging-interference or -exploitative competition occurs, thereby reducing food intake and potentially affecting health and reproduction, and (4) that human-induced alterations in food availability do not cause animals to modify their foraging behavior to the extent that natural history traits are altered and potentially maladaptive. As is apparent, animals must constantly trade off competing strategies to try to find the optimal solution to successfully survive and reproduce in their environment. Using a conservation behavior approach, we can understand such relationships that are critical to survival of individuals and persistence of populations.

2.3 Behavioral ecology - providing the mechanism

Behavioral ecology is a field of animal behavior that can be used to investigate fitness impacts of organismal interactions with their environment, since it seeks to understand both the ecological and evolutionary basis of animal behaviors. There are three fundamental types of adaptation that allow individuals to adjust to the environment: phenotypic plasticity, learning, and genetic (Huse and Giske 2004). These adaptations partly determine individual behavior, and whichever is dominant will depend on the current circumstances and the different timescales on which they function. Adaptation functions by animals making tradeoffs between competing goals to try and find an optimal solution that maximizes their fitness. Behavioral ecology therefore attempts to understand how an individual's behavior is adapted to the environment in which it lives, and how a particular behavior pattern contributes to an animal's chances of survival and its reproductive success (Krebs and Davies 1996). Furthermore, because anthropogenic change can disrupt optimal decision-making and affect an animal's reproductive success and survival, behavioral ecology can be a key ecological indicator when assessing wildlife fitness impacts. Within the field of behavioral ecology, there are three key behavior domains that are central to the attainment of high fitness in individuals of all species and are therefore of key concern in habitat-suitability and species-range effects management: foraging and predator–prey related behaviors, social behavior and reproduction, and life-history strategies (Caro 1998, Gill and Sutherland 2000, Festa-Bianchet and Apollonio 2003, Berger-Tal et al. 2011).

2.4 Spatial behavioral ecology - one step further

Because most wildlife management directives occur *in situ*, these domains are inherently related to spatiotemporal variations in landscape, and indeed, behavioral ecologists can benefit by assimilating the tools and the concepts developed in spatial ecology (Valcu and Kempenaers 2010). The following section focuses on how behavioral ecology combined with spatial ecology can be used to explain and explore space-use and movement patterns in wildlife.

2.4.1 Habitat selection

Conservation of a species requires knowledge of the habitat use of both sexes in order to predict the population size and to protect the habitats that a species requires. Habitat selection is the behavioral process used by individuals when choosing resources and habitats. From a behavioral ecology perspective, habitat selection implies that individual organisms have a choice of different types of habitat available to them, and that they actively move into, remain in, and/or return to certain areas over others (Stamps 2009). When faced with a site in which to forage, rest, or mate, an individual will rely on abiotic and biotic cues that will help shape the behavioral rules (optimal group size, anti-predator tradeoffs, foraging efficiency) and tactics (e.g., natal home range cues, public information cues and conspecific attraction) to make an optimal selection at various spatial and temporal scales (Johnson 1980).

Investigating habitat selection with a behavioral-ecological focus and using local, fine grain spatial parameters is common practice. However, more behavioral ecologists are availing themselves to the data-capturing tools and techniques offered by geographic information science, such as telemetry, remote sensing, and sensor networks, and incorporating larger-scale analyses to understand the complexities involved in animal habitat choice and use.

Indeed, behavioral ecology often contributes to habitat-selection studies and 'confounds' analyses relying just on empirical relationships between an organism and its static environment. Using GPS telemetry monitoring, Fischhoff et al. (2007) examined variation in plains zebra (*Equus burchelli*) movements and habitat use in relation to danger from lions. They found predator avoidance and predation risk to be the main drivers of habitat choice and movement patterns, and concluded that individual variation in zebra responses can affect individual variation in survival. Willems et al. (2009) used a remotely sensed index of plant productivity as a spatially explicit and temporally varying measure of habitat structure and productivity for the study of vervet monkey (*Chlorocebus pygerythrus*) habitat preferences. Using both broad spatiotemporal scales and finer grained level of analysis, they were able to relate home-range use to food availability, and anti-predatory responses to changes in habitat visibility using their index of vegetation productivity. Durães et al. (2007) evaluated whether female hot spots can account for patterns of lek structure in the blue-crowned manakin (*Lepidothrix coronata*) by modeling female distribution patterns relative to lek locations using radio-telemetry. The authors found a lack of spatial correlation between males and females, and concluded that refutation of the hotspot hypothesis renews the debate on how leks evolve and are shaped, and emphasizes that spatial considerations are an important issue for lek evolution that likely involve multiple interacting mechanisms. Lastly, using a combination of animal- and environmental-GPS point locations and satellite imagery, Greisser and Nystrand (2009) studied the influence of large-scale habitat structure on the vigilance levels of kin- and non-kin Siberian jay (*Perisoreus infaustus*) groups to aerial predators. They found that different foraging habitats, differentiated by large-scale metrics, had different levels of predation risk, and these were partially mediated by whether or not jays were in groups with offspring. The authors surmised that large-scale habitat structure influences predator–prey interactions; and therefore antipredator allocation is crucial to understanding spatial variation in habitat use and individual jay mortality. The above examples showcase the need of interrelating spatial data at fine and broad scales with fitness-maximizing behaviors and demonstrate the applicability of this approach in elucidating the array of factors involved in habitat use.

2.4.2 Dispersal and migration

Non-foraging movements of animals within a heterogeneous landscape are recognized as the key process influencing meta-population dynamics, the coexistence of competitors, community structure, disease ecology, and biological invasions (Morales and Ellner 2002). It is not surprising then, that most effort by conservationists has focused on the dispersal and migration requirements of animals. Animal dispersal consists of two component behaviors: (i) emigration out of an original habitat patch and (ii) subsequent search for a new habitat patch. Emigration is assumed to depend on the chance rate of encounter with habitat boundaries, and dispersers are assumed to search for new habitat in the manner of a correlated random walk (Conradt and Roper 2006). The decision-rules of animal movement, however, have a very strong behavioral component that is influenced by both endogenous and exogenous factors. Physiological and motivational states, perceived travel costs in terms of predation risk, and the distance at which a dispersing animal can perceive remote habitat will determine whether an animal will cross habitat gaps formed by fragmentation (Zollner and Lima 2005). Susceptibility to competition as well as level of conspecific attraction will also play an important role in determining the movements of individuals (Bélisle 2005).

Whether an animal needs to migrate to find resources or exploits resources from a central place to which it periodically returns will also affect the degree of impact from sub-optimal habitat quality, size, and connectivity. In other words, the movement paths of wildlife result from the dynamic interplay of the internal state of the organism, its motion capacity, its navigation capacity, and the external environment (Holyoak et al. 2008, Revilla and Wiegand 2008).

As with habitat-selection studies, behavioral ecologists also employ GIS techniques to both represent the environment and collect wildlife movement data when studying animal dispersal and migration. For instance, Long et al. (2008) investigated emigration cues and distance of transitional movements in white-tailed deer (*Odocoileus virginianus*), and found that both inbreeding avoidance and mate competition ultimately underlie emigration of juveniles, and that, proximately, these patterns of dispersal are elicited by different social cues during different seasons. Using ruffed grouse (*Bonasa umbellus*), Yoder et al. (2004) tested the hypothesis that increased movement rates during dispersal bouts increases conspicuousness and hence predation-related mortality of individuals. Contradictorily, they found that movement rates and distance moved did not predict bird mortality; instead, it was the familiarity with the site itself which determined the birds' survival. Lastly, a study by Hebblewhite and Merrill (2009) investigated how trade-offs between predation risk and forage differ between migrant strategies in migratory elk (*Cervus elaphus*). Each strategy had its associated costs and benefits, with resident elk balancing increased predation risk with refugia caused by human activities. These examples again highlight that the success of managers and policy makers when planning critical habitat for species conservation depends on a spatial *and* mechanistic understanding of the species in question.

2.5 Behavioral ecology and the individual

On a final note, it is crucial to realize that behavioral ecology concerns itself with the adaptations of *individuals*. Although inter-individual variation in phenotypic traits is omnipresent, it has historically been considered to be noise superimposed on the evolutionarily important signal, the population mean (Careau et al. 2008). But a rapidly growing literature on animal personality, temperament, coping styles, and behavioral syndromes (Stampes and Groothuis 2010) reveals the increasing importance researchers place on inter-individual variation as an important ecological and evolutionary characteristic of wild populations. Individuals are the building blocks of ecological systems - the birth and death of individuals are the constituents of the birth and death rates of populations, and because these rates are the result of the assimilated effects of varying and different fitness-maximizing behaviors that are used by each individual, population structure, demography, and community structure can be significantly affected by variation in the behavior of individuals (Bradbury *et al.* 2001). The approaches explained so far describe the relationships existing between a given individual organism, which is influenced by its need for basic resources (e.g. water, food, security cover, space), and the spatial distribution of such resources. As explained above, individuals are spatially clustered around resources and the spatial distribution of animals and plants can therefore be predicted. However, most animals and plants also have the need to encounter conspecifics and to reproduce. Therefore, populations are formed comprised of multiple individuals that are associated to spatially distributed resources and to each other, and these are the units that survive or go extinct during the evolutionary process. Subsequently, population-level properties such as persistence, resilience, and patterns of abundance over space and time are not simply the sum of the properties of individuals; instead, they emerge from the interactions of adaptive individuals with each other and with their environment (Figure 1). These links make models of spatial distribution of organisms and of populations relevant and crucial for the following conservation purposes: to predict spatial occurrence of populations, population sizes that resources can sustain, connectivity among populations, and their very chances of survival. As such, models of species distribution and habitat suitability should therefore consider individual mechanisms of habitat selection and movement coupled with spatially explicit representations of the animal's environment.

2.6 Behavioral ecology and SDMs

The call for integrating behavioral ecology into spatially explicit species distribution and range models is not new. Blumstein and Fernández-Juricic (2004) suggest that specific behavioral mechanisms should be the basis of bottom-up models that predict the behavior, movement, habitat use, and distribution of species of conservation concern. Morales and Ellner (2002) further posit that the challenge for scaling up movement patterns resides in the complexities of individual behavior, specifically behavioral variability between individuals and within an individual over time, rather than solely in the spatial structure of the landscape. Bélisle (2005) also advocates for the use of behavioral ecological resource-based

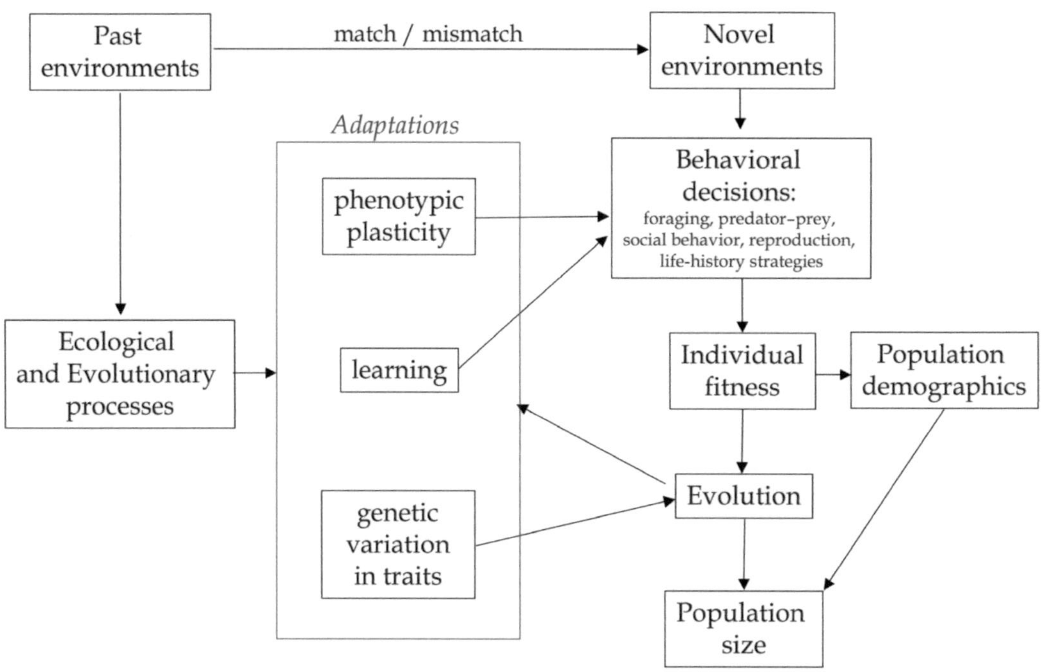

Fig. 1. Summary of how behavioral decisions, driven by animal adaptations that have evolved over time from ecological and evolutionary processes, can match or mismatch an animal to its environment. This process has cascading direct and indirect effects on population and species persistence via individual-fitness effects on population demographics and evolution of species' traits. Modified from Lankau et al. (2011).

models in judging habitat quality, travel costs, and hence landscape functional connectivity. Specifically, these latter types of models would be capable of addressing the distribution of individuals among resource patches at large spatial scales, among resource patches embedded within a hierarchy of spatial scales, and along smoothly changing resource gradients. Finally, Jonzén (2008) acknowledges that while habitat selection theory has a successful history in behavioral ecology, it can also be useful for understanding spatial population dynamics on a large scale. We propose here that the principles of behavioral ecology can be quite naturally and readily integrated with the tenets of spatial ecology in the alternative approach to SDM:

3. Agent-based models

Agent-based models (ABMs) are computational simulation tools that rely on a bottom-up approach that explicitly considers the components of a system (i.e. individual entities represented as agents) and attempts to understand how the system's properties emerge from the interactions among these components (Grimm 1999, Grimm and Railsback 2005). This emphasis on interactions between agents and their environment is what distinguishes agent-based modeling (also referred to as individual-based models) from other systemic modeling approaches (Marceau 2008; Figure 2a), and additionally allows the use of ABMs for the exploration of complex phenomena that are ill-suited to analytic approaches (e.g., statistical models; Tang 2008).

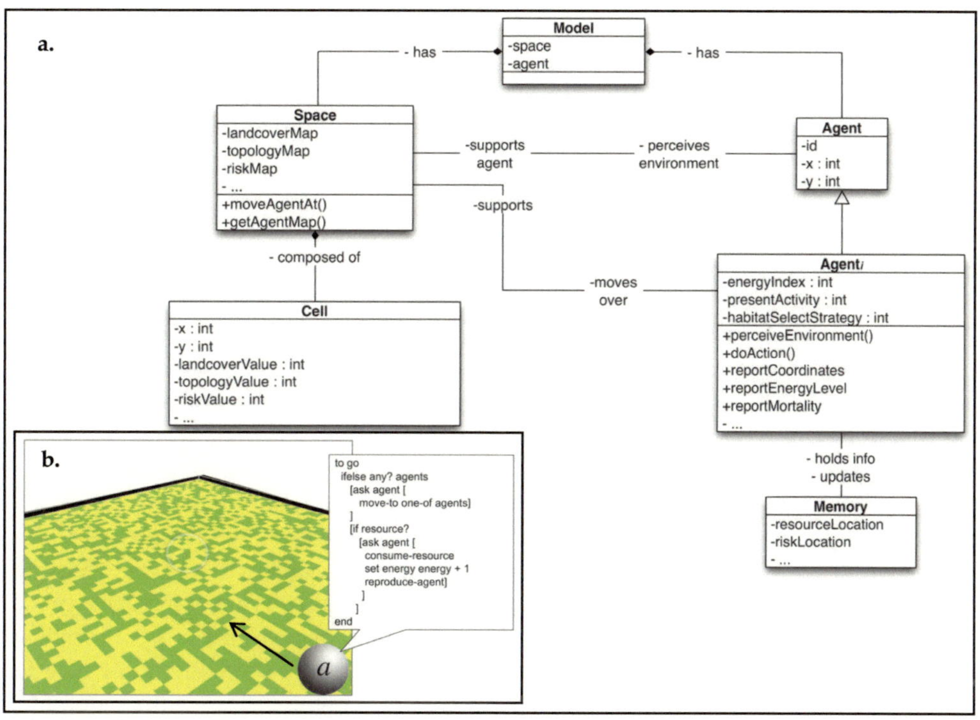

Fig. 2. ABM architecture. a) Example of an ABM conceptual diagram demonstrating how agents are tightly coupled to their environment. b) Generic programming language, giving rise to agent (*a*) autonomy and intelligence.

The concepts underlying ABM are similar to those of the object-oriented programming paradigm in computer science, and ABMs frequently employ object-oriented programming languages like C++ and Java (An et al. 2005; Figure 2b). Because of this architecture, the most critical feature of ABMs is their ability to reproduce artificial intelligence. Agents can explicitly execute decision-making heuristics - symbolic rules or numerical functions - that can be either predefined (e.g., expert knowledge or statistical inferences) or learned through their interactions and feedback with other agents or their environment (e.g., via memory or machine learning techniques like genetic and evolutionary algorithms; Russell and Norvig 1995, Tang 2008). These agents act independently of any controlling intelligence, they are goal-driven and try to fulfill specific objectives, they are aware of and can respond to changes in their environment, they can move within that environment, and they can be designed to learn and adapt their state and behavior in response to stimuli from other agents and their surroundings. It is these characteristics of ABMs that make their amalgamation with animal mechanisms of habitat selection and movement so ideal as they share the same principles of behavioral ecology: animal adaptation, individual variation, and fitness-maximizing tradeoff behaviors.

3.1 Behavioral-ecological ABMs and species distributions

ABMs have been developed to expressly evaluate wildlife habitat suitability and species range effects via habitat-selection and movement studies. These ABMs can be divided into categories depending on whether agents are given imposed, empirically-derived behaviors, or agents are allowed to choose the optimal strategy themselves based on decision-making tradeoffs (for a thorough review, see McLane et al. 2011). The latter category is the focus of this section, as it most closely represents the tenets of behavioral ecology (Figure 3). As one example of habitat suitability and its underlying habitat-selection behaviors, Kanarek et al. (2008) incorporated habitat selection in their ABM of environmental fluctuations on a barnacle geese (*Branta leucopsis*) population in Helgeland, Norway. The aim of each individual was to optimize fitness (survival and reproduction) by gaining enough food (energy reserves) to meet a threshold of energy necessary for successful reproduction. In their model, geese chose unoccupied habitat according to their rank in the population-structured dominance hierarchy, their memory of previously visited sites in past years, past reproductive success, inherited genetic influence towards site preference, and knowledge of the available biomass density. Their findings revealed that different types of population dynamics and patterns of colonization occur, depending on the strength of site fidelity and degree of habitat loss. Duriez et al. (2009) investigated the decision rules of departure and stopover ecology of the migratory behavior of geese (*Anser brachyrhynchus*) between wintering grounds in Denmark and breeding grounds in Svalbard, Norway. They tested rules governed by energetics, time-related cues and external cues by comparing predicted and observed departure dates. The most accurate predictions were made by a combination of cues including: the amount of body stores, date, and plant phenology. They also found that by changing decision rules over the course of the migration, with external cues becoming decreasingly important and time-related cues becoming increasingly important as the geese approached their breeding grounds, they could improve ABM model predictions of site selection.

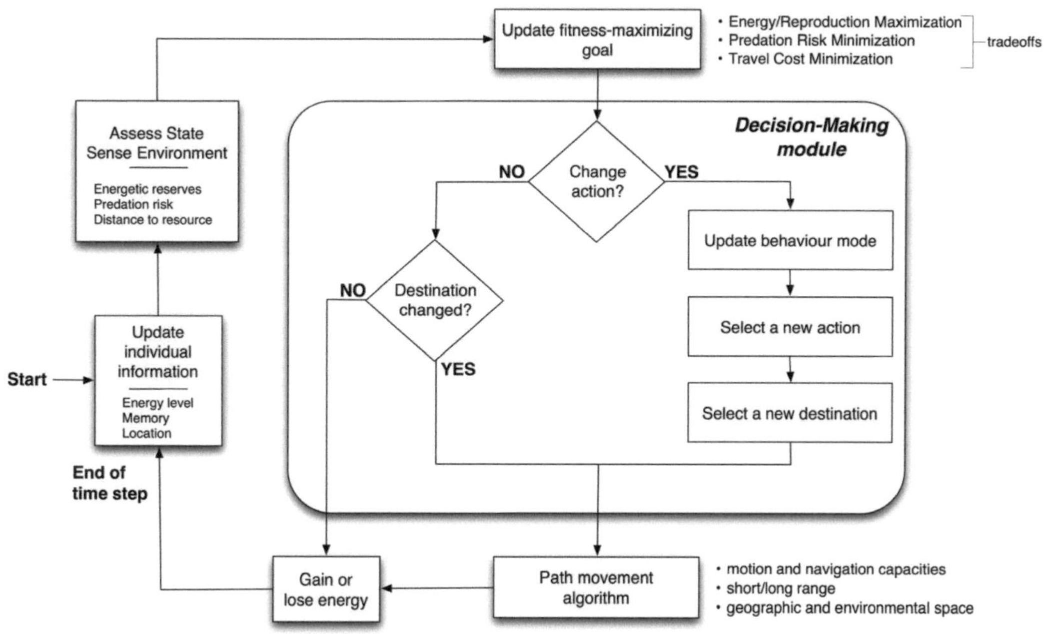

Fig. 3. An agent's decision-making heuristics in a behavioral-ecological ABM.

With respect to range-limiting effects and migration, Pettifor et al. (2000) used an agent-based approach to predict the response of goose populations to both natural and human-induced environmental changes. They used contrasting time-minimizing vs. energy-maximizing foraging strategies as well as a game theoretic approach of competitor density to determine year-round dynamics of the goose populations. Populations were predicted to decline following habitat loss in their winter or spring-staging sites, providing a clear illustration of the need for a year-round, individual-behavior approach to animal population dynamics. Lastly, Goss-Custard and Stillman's (2008) seminal work on oystercatcher (*Haematopus ostralegus*) management elegantly demonstrates how mechanistic ABMs can contribute to the conservation of local populations' occupancy and species persistence. The overall purpose of their ABM was to predict how environmental change (e.g., habitat loss, changes in human disturbance, climate change, mitigation measures in compensation for developments, and changes in population size itself) affects the survival rate and body condition in animal populations. The model does this by predicting how individual animals respond to environmental change by altering their feeding location, consuming different food or adjusting the amount of time spent feeding. The central assumption of the model is that animals behave in ways that maximize their chances of survival by using rate-maximizing optimization decision rules and game theoretic rules in that each animal responds to the decisions made by competitors in deciding when, where, and on what to feed. They found that even small reductions in fitness can substantially reduce population size of shorebirds since their "ecological food requirement" greatly exceeds the "physiological requirement".

As has been demonstrated, the use of behavioral-ecological based ABMs can produce emergent system-level processes that allow one to ask ecological questions that extend beyond the individual itself. Imposing system behavior by giving individuals mechanical, empirically-derived traits can also provide a feasible alternative. However, this might lead to the simple reproduction of reactive abilities and behaviors observed in real systems without providing the desired ultimate causations necessary to understand animal movements and habitat selection. This distinction is particularly important for wildlife management such as ecological forecasting. In fact, SDM approaches may not reliably ascertain whether the empirical relationship upon which these models are based will hold under new environmental conditions

To have confidence in predictions, models need to operate on basic principles, underpinned by theory that will still apply in the new scenarios, rather than on present-day empirical relationships which may no longer hold in the scenarios for which predictions are required (Grimm et al. 2007). The allocation of behavioral strategies to individual agents allows researchers to predict how animals will most likely respond to novel changes in their environment, since the underlying processes are consistent with evolutionary concepts (i.e., how animals will tradeoff fitness-maximizing behaviors and find an optimum). Finally, with ABMs intra-specific relationships among individuals can also be modeled, thus allowing better understanding of population responses to the environment and to conspecifics as well as other organisms (e.g. competitors, predators, parasites).

4. ABMs and issues of scale

All types of animal-environment models need to allow for the determination of where the important interactions lie and to understand both the spatial scales and time scales on which the various processes operate (Bithell and Brasington 2009). This is particularly the case where the issues of conservation planning and ecological forecasting are concerned, as these typically involve spatial scales that can cross political borders, temporal scales longer than the organism's lifespan, and the need for long-term institutional policies to be effective. Because the dynamic nature of the environment plays such an influential role in affecting organism state, behavioral decisions and motion, a representation of the animal's actual environment in a spatially explicit manner at the adequate spatial and temporal scale can improve the effectiveness of wildlife management as it can highlight the causal links between organism movement and environmental change (Nathan et al. 2008; Figure 4).

4.1 Extent versus resolution

Although various approaches exist, there is as yet little consensus on how to deal with scale disparities - such as extent and resolution, when fitting SDMs (Barry and Elith 2006; Elith and Leathwick 2009). While there is no single scale at which ecological patterns should be studied (Levin 1992), mismatches between coverage and grain can be caused by the study goals, the system, data availability, and by extent to which a species perceives its environment. Some SDMs attempt to address these issues by incorporating hierarchical structures into the modelling process, either through the use of sub-models, through Bayesian approaches that operate across scales, or through models that allow nested structures of data (reviewed in Elith and Leathwick 2009). However, these different approaches remain untested both theoretically and practically, nor is it certain whether these scale-specific model predictors provide a clear advantage over traditional SDMs (Barry and Elith 2006).

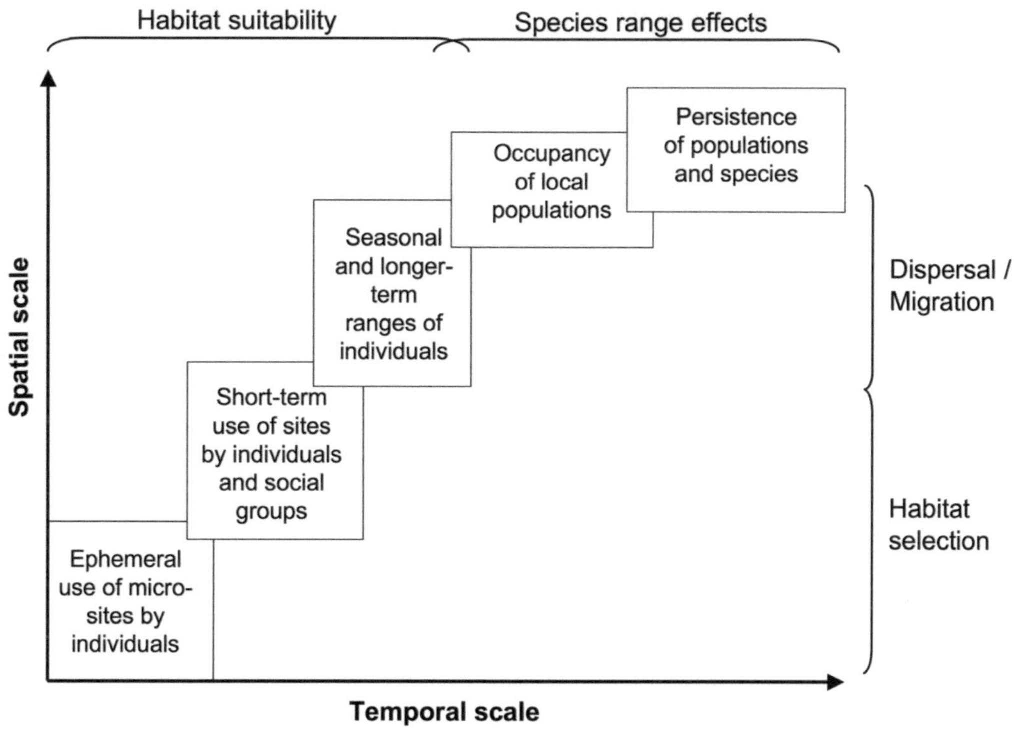

Fig. 4. The link between spatial and temporal scales of habitat selection and dispersal/ migration and conservation planning of habitat suitability and species range effects. Modified from Mayor et al. (2009).

Agent-based models are particularly well suited to represent a virtual geographic environment within which entities and their interrelationships (e.g., spatial, temporal, and spatiotemporal) can be explicitly described, and provide contextual information to which agents sense and respond (Tang 2008). In agent-based modelling, the movement trajectory or pathway of an animal can be represented as a sequence of discrete time-stamped location variables, for example, geographic coordinates. Because environment representation in ABMs can be raster- or vector-based, the location variables can be further indexed by raster cells or vector-based patches (Tang and Bennett 2010; McLane et al. 2011). ABMs are not completely immune to issues of scale. Scale factors can affect the design and application of agent-based models particularly when temporal landcover changes are incorporated. To deal with spatial constraints, Evans and Kelley (2004) recommend that models be run at a range of spatial scales. Then modelers can choose the minimally-acceptable resolution by identifying the spatial resolution at which agents have sufficient partitions on their landscapes within which to make biologically-relevant decisions pertinent to the study goals (minimum change unit), and where the heterogeneity of the landcover and land suitability measures are adequately represented. The coarsest, or upper bound, resolution for model runs can be identified by the resolution at which appreciable data loss occurs (e.g., the disappearance of potentially relevant cover classes).

Despite the universal confounds of scale regardless of the modelling methodology used, ABMs are still more decoupled from scale issues than SDMs as the researcher can address the extent- and resolution-issues by developing a model that makes the best statistical use of information at the finest spatial and temporal resolution available; and then allowing large-scale behavior to emerge from the small scale via interaction between these model elements (Parker et al. 2003; Bithell and Brasington 2009). In addition, because ABMs incorporate ecological theory, and deal with processes and mechanisms at the level of the individual, the resultant hierarchical phenomena that emerge from agents' interactions with others and their environment can naturally accommodate issues of scale (Breckling et al. 2006). As an illustrative example, Bennett and Tang (2006) combined cell- and patch-based approaches to represent multi-scale environmental representation in their elk migration model. Agent elk performed local movement at the cell level, but were capable of perceiving and using greater scale, patch-level information to guide their long-distance winter migration. In the wolf (*Canid lupus*) ABM study of Musiani et al. (2010), their canid agents were able to perceive disturbance (i.e., bear and human agents) at a 200m scale, and able to detect prey (elk) at a 3km scale, travel accordingly, and allow pack home range dynamics to emerge from these interactions and behaviors.

Multiscale detection does not have to only be via the the agent's immediate perception of heterogeneous landscapes features and/or agents at different scales, but through its memory processes. In an ABM study of the effect of anthropogenic landscape change on disease of red colobus monkeys (*Procolobus rufomitratus*) populations (Bonnell et al. 2010), monkey agents were able to remember the location and quantity of past resource sites that contained a significantly higher amount of resources (i.e., spatial memory), allowing red colobus agents to estimate resource levels at these sites while not within their search radius. This allowed for a more biologically-relevant prediction of the optimal distribution of resources which could facilitate the spread of an infectious agent through the simulated population. ABMs, through a multi-scale environmental representation, can therefore support the investigation of scale issues and even facilitate our understanding of individual movement behavior in response to spatiotemporal heterogeneity on landscapes in ways in which traditional SDMs cannot (Figure 5).

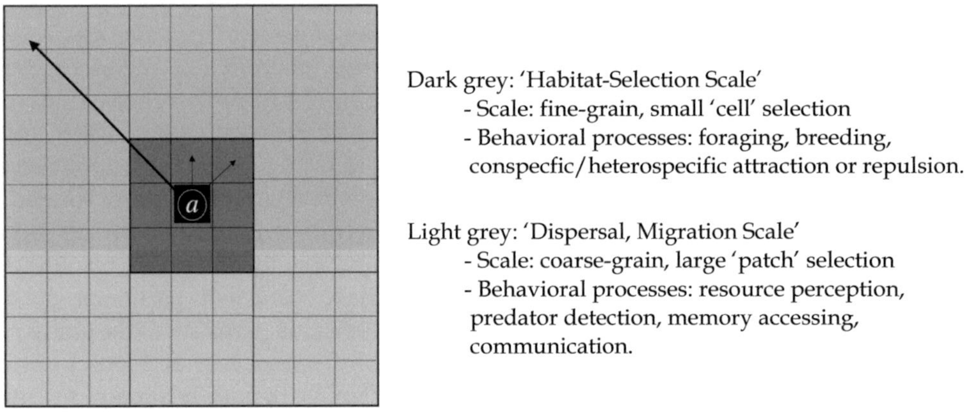

Fig. 5. Multiscale habitat selection of an agent (*a*) in a spatially-explicit space.

4.2 Geographic versus environmental space

Another issue in SDMs is the distinction between geographic and environmental space. For example, two animal locations may be very close in geographic space, but the two points may be in completely different habitats. Important geographic predictors include glaciation, fire, contagious diseases, and connectivity (Elith and Leathwick 2009). Environmental factors primarily deal with abiotic and biotic processes such as resource distribution, social factors, and predation risk. Purely geographic SDMs, when attempting to derive habitat suitability and extrapolate findings to predictive species-range modeling, may ignore important environmental predictors. Equally, SDMs that solely incorporate environmental variables have difficulty in mapping their predictions onto geographic space as species distribution simply reflects the spatial autocorrelation of the environment. Current methods using both geographic and environmental predictors in SDMs (examples include species prevalence, latitudinal range / marginality, and spatial auto-correlation), while a promising compromise, can affect modelling performance and species predictions, with contradictory results (Marmion et al. 2009). Furthermore, these combined-effects models are more difficult to implement than standard techniques so they are under-utilized, and the emerging recommendation is to simultaneously apply several SDM methods within a consensus modelling framework (Grenouillet et al. 2011).

ABMs are capable of representing both geographic and environmental space cohesively. This is accomplished by coupling ABMs to geographic information systems (GIS) that provide detailed abiotic and biotic characteristics of the environment (e.g., land cover, elevation models, resource distributions, risk), and having agents assign values to these geographic and environmental attributes either via a weighting function (like a friction map) or independently (Brown et al. 2005; Figure 6). The decision-making behaviors of agents therefore consider the spatiotemporal variation of the landscape itself; and the ABM accommodates how this variation feedbacks onto behavior in dynamic, non-predictable and non-linear ways. Specifically, an animal's location in space and time, the way it perceives the surrounding landscape, and its subsequent behavior all determine what resources are accessible to it and what it chooses among those resources (May et al. 2010). In ABMs, the scale and degree of heterogeneity within the landscape will be perceived in different ways by different species, and thus an animal's perception will influence its movement behavior, choice of search strategy and habitat patch choice (e.g. Lima and Zollner, 1996).

In essence, by allowing agents to explicitly interact with, modify, and respond to their environs, geographic and environmental predictors are both naturally incorporated into the agent's decision-making process. Any habitat-selection or movement patterns that then emerge will be more robust to the uncertainties involved in future predictions of species occupancy and range effects since specific geographical factors (e.g., barriers to movement, events) and spatial autocorrelation are directly represented and assimilated into the model. As an illustrative example, Rands et al. (2004) created a state-dependent foraging ABM for social animals in selfish (i.e., non-kin) herds. In the model, the agents tradeoff protective herding versus individual foraging behavior, with the individual basing its decisions upon its energy reserves, the distribution of foraging resources in the environment, and the perceptual range over which individuals are able to detect conspecifics, risks, and resources. The resulting behavior and energetic reserves of individuals, and the resulting group sizes were shown to be affected both by the ability of the forager to detect conspecifics and areas of the environment suitable for foraging, and by the distribution of energy in the environment. Both environmental (presence of conspecifics) and geographic (spatial detection of resources) are considered independently of one another with this model. Grosman et al. (2009) developed an ABM to investigate management strategies that would reduce moose-vehicle collisions through salt-pool removal and displacement. The moose agents forage and travel in the Laurentides Wildlife Reserve, Quebec; and assess patches to visit and disperse through based on a weighted assessment of both geographic and environmental factors of food quality, cover quality (protection from predators and thermal stress), proximity to salt pools, proximity to water, and slope. The realistic patterns which emerged from the simulations revealed that the most successful management action was complete removal of salt-pools without any compensatory ones to ensure moose (*Alces alces*) survival.

The ABM examples used in this section either comprised behavioral mechanisms in a spatially-implicit environment, or incorporated and modeled empirically-driven behaviors of agents (e.g., probabilistic, mechanical 'decision-making') on spatially realistic landscapes. Each proved very capable of accommodating multiscale agent behaviors and multi-environmental factors in reproducing the desired results. We believe, however, that integrating multi-scale and -environs using more behavioral-ecological based mechanisms in spatially realistic contexts (of which explicit examples in the literature are not yet available) will prove to be even more beneficial. When combined with behavioral mechanisms, the realism and applicability of the model will increase multi-fold, and the capacity of these ABMs to accommodate the dynamism of the environment, the spatial patterns of inter- and intra-species mechanisms, and the feedbacks and adaptations inherent in these systems will represent a powerful tool in conservation planning and ecological forecasting.

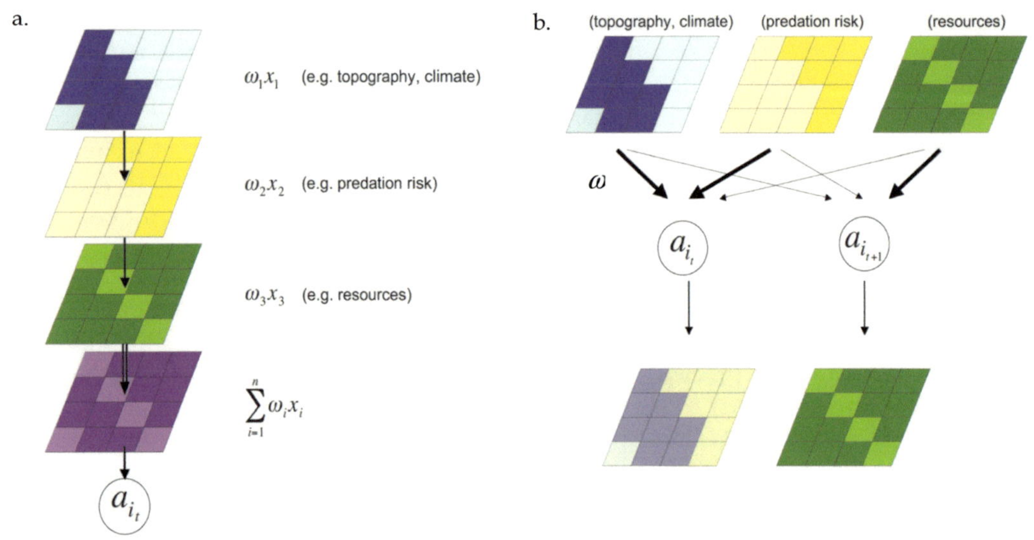

Fig. 6. Examples of how an agent perceives its environment. a.) Geographic and environmental variables are given specific weights (*w*) based on the agent's habitat preferences at the initialization of the model, and the agent assesses its environs based on the integration of these factors which remain unchanged throughout the simulation. b.) Geographic and environmental variables are independently assessed by the agent at each time step, and factors are weighted based on the agent's internal state, and/or fitness-maximizing goal at that time step.

5. ABMs and model evaluation

Because SDMs are essentially statistical models, it is necessary to consider the compatibility between statistical model evaluation and selection of predicted versus observed species distributions and the underlying ecological model. The principle of model selection is to formulate different verbal hypotheses, express these hypotheses mathematically as statistical models, evaluate a score of a goodness-of-fit indicator for each statistical model, and either strongly select one hypothesis or keep a set of plausible ones with different weights (Piou et al. 2009). There still exists a lack of agreement amongst SDM researchers about the most effective statistical methods to evaluate and predict the spatial distribution and habitat selection of animal species, and the degree of ecological realism inherent in the statistically 'best-fit' model (Keating and Cherry 2004; Guisan and Thuille 2005; Austin 2007; Elith and Graham 2009). The main reason is that different statistics are used for the various different models, each one measuring different aspects of performance, and as such, appropriate statistics relevant to the application of the model need to be selected.

ABMs use an altogether different approach, known as pattern-oriented modeling (POM). This protocol is based on the assumption that patterns are the defining characteristics of a system and are indicators of essential underlying ecological structures and processes. Patterns are defined by Grimm et al. (2005) as any observation made at any hierarchical level or scale of the real system that display non-random structure. Patterns are therefore particular expressions of a given comportment of the studied individuals, populations, or system. POM requires the researcher to begin with a pattern found in the real system, posit hypotheses to explain the pattern, and then develop predictions which can be tested. By observing multiple patterns at different hierarchical levels and scales, one can systematically optimize model complexity, parameterize the model, and simultaneously make it more general and testable (Grimm et al. 2005).

POM capitalizes on both behavioral ecology and spatial ecology through the emergence of biologically- (and behaviorally-) relevant patterns at multiple scales to evaluate model results. For example, the emergence of a pattern generated by a tradeoff between the costs and benefits of a decision process could explain the selection pattern of certain habitats, leading to specific step length and turning angle distribution patterns, and allowing the reproduction of home range characteristics to emerge (Latombe et al. 2011). This modelling approach allows one to simultaneously filter combinations of parameter values and model structures in order to achieve the aims of testing the behavior of the agents in the model and of reducing parameter uncertainty. The greater the number of real-world patterns that can be simulated concurrently, the greater the confidence in the model, and typically the smaller the possible parameter space (Topping et al. 2009). By extension, the POM approach can additionally allow for rigorous statistical approaches. Information theory and information criteria have been recently developed for the POM method, and serves to further improve the agent-based modeling framework (Piou et al. 2009). The approach can be used to analyze separately the different patterns of focus, and analyze together an overall level of evidence of each model to all the patterns. This approach is more universal than the various methods of SDMs model evaluation, and can be applied to very different types of agent-based models.

POM has been used extensively and demonstrates a strong utility in addressing model complexity, unknown data requirements, variable parameterization, and model evaluation. As an example, Railsback and Harvey (2002) created an ABM to simulate habitat selection of salmonid fish species in response to spatial and temporal variation in mortality risks and food availability. They used their ABM to draw conclusions about foraging theory by analyzing the ABM's ability to reproduce six patterns of habitat selection by contrasting three alternative habitat-selection objectives: maximizing current growth rate, current survival probability, or expected maturity. In the model, fish based their daily decision on the projection of current habitat conditions for a certain number of days into the future, as this strategy was capable of reproducing a set of six patterns observed in reality. Rossmanith et al. (2006) developed an ABM to test the impact of three behavioral scenarios on population persistence of the lesser spotted woodpecker *Picoides minor*: strict monogamy, polyandry without costs, and polyandry assuming costs in terms of lower survival and reproductive success for secondary males. Using a POM-approach where the model was simultaneously fitted to a set of four empirically observed patterns (adult sex ratio, ratio of old and new pairs, proportion of nest producing at least one fledgling, number of fledglings per successful nest) to produce a realistic population structure, the authors found that polyandry and in general flexibility in mating systems is a buffer mechanism that can significantly reduce the impact of environmental and demographic fluctuations that cause variations in the population's growth rate. Consequently, they suggested that rare, exceptional behavior should be considered explicitly when predicting the persistence of populations. Lastly, Tyre et al. (2007) explored behavioral mechanisms for home range overlap in a Scincid lizard, *Tiliqua rugosa*. The authors tested two mechanisms, one that used refuge sites randomly and one that included a behavioral component that incorporated refuge sites based on nearest neighbor distances and use by conspecifics. Comparisons between the simulated patterns and the observed patterns of range overlap provided evidence that the behaviorally-driven refuge use model was a better approximation of lizard space use. In sum, pattern-oriented modelling presents an effective method for identifying and evaluating behavioral mechanisms of habitat selection and animal movement underlying observed patterns.

6. Conclusion

In a recent paper, Caro and Sherman (2011) state that the field of behavioral ecology is at a key turning point in its history. While the discipline was originally created with the intent of developing explanatory theories of ecological and evolutionary adaptations of organisms, future studies should be designed to provide information for the protection and management of organisms that are increasingly being compromised in human-dominated landscapes because of species extinctions, habitat destruction, invasive species, pollution, and climate change. The authors posit that behavioral ecology and conservation biology can be linked by forecasting how anthropogenic ecological changes are liable to reshape specific aspects of behavioral ecology during the 21st Century. We would like to further add that Caro and Sherman's 'call to arms' can be accomplished in one manner by integrating behavioral ecology with spatial ecology in agent-based models for conservation planning.

As we have shown, ABMs have multiple advantages: they incorporate and embody individual variation, adaptation, emergence from interactions, geographic and environmental space, short- and long-range spatial scales, multiple processes, and hypothesis testing to identify the most influential mechanisms. In doing so, ABMs can reduce uncertainty and increase model fit in the identification of habitat suitability and in the prediction of long-term species responses to environmental change. In addition, ABMs are ideally suited to work across spatial and temporal scales and on individuals and populations of organisms, thus reaching the most meaningful scale in conservation biology. ABMs also have the ability to incorporate dynamic interactions between individuals, whether they be competitors, predators, or even humans (e.g., hunters, recreationists). Since the models are not constructed to meet a set of equilibrium criteria, they can additionally produce discontinuous and nonlinear phenomena, such as species extinctions, range shifts, and exponential growth or decline of populations (Parker et al. 2003). And to reiterate, employing behavioral ecological concepts to reproduce the underlying mechanisms can aid in overcoming the issues typically associated with traditional SDMs.

Our intent here is not to suggest ABMs replace statistical SDMs. They simply represent a promising alternative approach. Spatially-explicit, behavioral-ecological based ABMs are still rare; most models found in the literature are empirical and/or are based in implicitly-structured spatial environs (see McLane et al. 2011 for a review). ABMs also need more testing and comparisons, of their own predictions and with those of other models, although there has been recent progress in this regard (Latombe et al. 2011). Nonetheless, while we perceive ABMs that encompass such a multidisciplinary approach as promising species distribution models for conservation research, the full potential of agent-based modeling in this domain still remains to be explored and fulfilled.

7. References

An, L., Linderman, M., Qi, J., Shortridge, A., and Liu, L. 2005. Exploring complexity in a human-environment system: an agent-based spatial model for multidisciplinary and multiscale integration. *Annals of the Association of American Geographers* 95:54–79.

Austin, M. 2007. Species distribution models and ecological theory: A critical assessment and some possible new approaches. *Ecological Modelling* 200:1-19.

Barry, S.C., and Elith, J. 2006. Error and uncertainty in habitat models. *Journal of Applied Ecology* 43:413–23.

Bastille-Rousseau, G., Fortin, D., and Dussault, C. 2010. Inference from habitat-selection analysis depends on foraging strategies. *Journal of Animal Ecology* 79:1157–1163.

Bélisle, M. 2005. Measuring landscape connectivity: The challenge of behavioral landscape ecology. *Ecology* 86:1988–1995.

Bennett, D.A., and Tang, W. 2006. Modelling adaptive, spatially-aware, and mobile agents: elk migration in Yellowstone. *International Journal of Geographic Information Science* 20:1039–1066.

Berger-Tal, O., Polak, T., Oron, A., Lubin, Y., Kotler, B.P., and Saltz, D. 2011. Integrating animal behavior and conservation biology: a conceptual framework. *Behavioral Ecology* 22(2):236-239.

Beyer, H.L., Haydon, D.T., Morales, J.M., Frair, J.L., Hebblewhite, M., Mitchell, M., and Matthiopoulos. 2010. The interpretation of habitat preference metrics under use-availability designs. *Philosophical Transactions of the Royal Society B* 365:2245-2254.

Bithell, M., and Brasington, J. 2009. Coupling agent-based models of subsistence farming with individual-based forest models and dynamic models of water distribution. *Environmental Modelling and Software* 24(2):173–190.

Blumstein, D. T., and Fernández-Juricic, E. 2004. The emergence of conservation behavior. *Conservation Biology* 18:1175–1177.

Bonnell, T.R., Sengupta, R.R., Chapman, C.A., and Goldberg, T.L. 2010. An agent-based model of red colobus resources and disease dynamics implicates key resource sites as hot spots of disease transmission. *Ecological Modelling* 221:2491-2500.

Bradbury, R.B., Payne, R.J.H., Wilson, J.D., and Krebs, J.R. 2001. Predicting population response to resource management. *Trends in Ecology and Evolution* 16:440–445.

Breckling, B., Middelhoff, U., and Reuter, R. 2006. Individual-based models as tools for ecological theory and application: understanding the emergence of organizational properties in ecological systems. *Ecological Modelling* 194:102–113.

Brown, D.G. 2005. Agent-based models. In: Geist, H. (Ed.), *Our Earth's Changing Land: An Encyclopedia of Land-Use and Land-Cover Change*. Greenwood Press, Portsmouth, NH.

Buchholz, R. 2007. Behavioral biology: an effective and relevant conservation tool. *Trends in Ecology and Evolution* 22:401-407.

Buckley, L.B., Urban, M.C., Angilletta, M.J., Crozier, L.G., Rissler, L.J., and Sears, M.W. 2010. Can mechanism inform species' distribution models? *Ecology Letters* 13:1041–1054.

Careau, V., Thomas, D., Humphries, M.M., and Reale, D. 2008. Energy metabolism and animal personality. *Oikos* 117:641–653.

Caro, T.M. 1998. Editor, *Behavioral Ecology and Conservation Biology*. Oxford University Press, UK.

Caro, T.M., and Sherman, P.W. 2011. Endangered species and a threatened discipline: behavioral ecology. *Trends in Ecology and Evolution* 26(3):111-118.

Conradt, L., and Roper, T.J. 2006. Nonrandom movement behavior at habitat boundaries in two butterfly species: Implications for dispersal. *Ecology* 87:125–132.

Duckworth, R.A., and Badyaev, A.V. 2007. Coupling of dispersal and aggression facilitates the rapid range expansion of a passerine bird. *Proceedings of the National Academy of Sciences of the United States of America* 104:15017–15022.

Durães R., Loiselle B.A., and Blake J.G. 2007. Intersexual spatial relationships in a lekking species: blue-crowned manakins and female hot spots. *Behavioral Ecology* 18:1029–1039.

Duriez, O., Baure, S., Destin, A., Madsen, J., Nolet, B.A., Stillman, R., and Klaassen, M. 2009. What decision-rues might pink-footed geese use to depart on migration? An individual-based model. Behavioral *Ecology* 20:560–569.

Elith, J., and Graham, C. 2009. Do they? How do they? WHY do they differ? On finding reasons for differing performances of species distribution models. *Ecography* 32:66–77.

Elith, J., and Leathwick, J.R. 2009. Species distribution models: ecological explanation and prediction across space and time. *Annual Review of Ecology, Evolution and Systematics* 40:677-697.

Elith, J., Kearney, M., and Phillips, S. 2010. The art of modeling range-shifting species. *Methods in Ecology and Evolution* 1:330-342.

Evans, T. P., and H. Kelley. 2004. Multi-scale analysis of a household level agent-based model of landcover change. *Journal of Environmental Management* 72:57-72.

Festa-Bianchet, M., and Apollonio, M. 2003. *Animal behaviour and wildlife conservation*. Washington: Island Press

Fieberg, J., Matthiopoulos, J., Hebblewhite, M., Boyce, M. S., and Frair, J. L. 2010 Correlation and studies of habitat selection: problem, red herring or opportunity? *Philosophical Transactions of the Royal Society B* 365:2233–2244.

Fischhoff, I.R., Sundarasen, S.R., Cordingly, J., and Rubenstein, DI. 2007. Habitat use and movements of plains zebra (*Equus burchelli*) in response to predation danger from lions. *Behavioral Ecology* 18(4): 725-729.

Fretwell, S. D., and Lucas, H. L., Jr. 1969. On territorial behavior and other factors influencing habitat distribution in birds. I. Theoretical Development. *Acta Biotheoretica* 19:16–36.

Gill, J.A., and Sutherland, W.J. 2000. The role of behavioral decision-making in predicting the consequences of human disturbance. In: Gosling, L.M., and Sutherland, W.J. (Eds.), *Behaviour and Conservation.* Cambridge University Press, Cambridge.

Goss-Custard, J.D., and Stillman, R. 2008. Individual-based models and the management of shorebird populations. *Natural Resource Modelling* 21:3–71.

Grenouillet, G., Buisson, L., Casajus, N., and Lek, S. 2011. Ensemble modelling of species distribution: the effects of geographical and environmental ranges. *Ecography* 34:9-17.

Griesser, M., and Nystrand, M. 2009. Vigilance and predation of a forest-living bird species depend on large-scale habitat structure. *Behavioral Ecology* 20:709-715.

Grimm, V. 1999. Ten years of individual-based modeling in ecology: what have we learned and what could we learn in the future? *Ecological Modelling* 115, 129–148.

Grimm, V., and Railsback, S.F. 2005. *Individual-based Modeling and Ecology.* Princeton University Press, Princeton, New Jersey.

Grimm, V., Revilla, E., Berger, U., Jeltsch, F., Mooij, W.M., Railsback, S.F., Thulke, H.H., Weiner, J., Wiegand, T., and DeAngelis, D.L. 2005. Pattern-oriented modeling of agent based complex systems: lessons from ecology. *Science* 310:987–991.

Grimm, V., Stillman, R., Jax, K., and Goss-Custard, J. 2007. Modeling adaptive behavior in event-driven environments: temporally explicit Individual-based Ecology. In: Bissonette, J., Storch, I. (Eds.), *Temporal Dimemsions of Wildlife Ecology: Wildlife Responses to Variable Resources.* Springer, pp. 59-73.

Grosman, P.D., Jaeger, J.A.G., Biron, P.M., Dussault, C., and Ouellet, J.P. 2009. Reducing moose–vehicle collisions through salt pool removal and displacement: an agent based modeling approach. *Ecology and Society* 14:17. URL: http://www.ecologyandsociety.org/vol14/iss2/art17/

Guisan, A., and Thuille, W. 2005. Predicting species distribution: offering more than simple habitat models. *Ecology Letters* 8: 993–1009.

Guisan, A., and Zimmermann, N.E. 2000. Predictive habitat distribution models in ecology. *Ecological Modelling* 135:147–186.

Hebblewhite, M., and Merrill, E. H. 2009 Trade-offs between predation risk and forage differ between migrant strategies in a migratory ungulate. *Ecology* 90:3445–3454.

Holyoak, M., Casagrandi, R., Nathan, R., Revilla, E., and Spiegel, O. 2008, Trends and missing parts in the study of movement ecology. *Proceedings of the National Academy of Sciences* 105: 19060-19065.

Huse, G., and Giske, J. 2004. Scales of adaptation in Individual-based modeling. In: Seuront, L., and Strutton, P.G. (Eds.), *Scales in aquatic ecology. Measurement, analysis, simulation.* CRC Press p 507-521.

Johnson, D.H. 1980. The comparison of usage and availability measurements for evaluating resource preference. *Ecology* 61:65–71.

Jonzén, N. 2008. Habitat selection: implications for monitoring, management, and conservation. *Israel Journal of Ecology & Evolution* 54:459-471.

Kanarek, A.R., Lamberson, R.H., and Black, J.M. 2008. An individual-based model for traditional foraging behavior: investigating effects of environmental fluctuation. *Natural Resource Modeling* 21:93–116.

Kearney, M., and Porter, W.P. 2009. Mechanistic niche modelling: combining physiological and spatial data to predict species' ranges. *Ecology Letters* 12:334–350.

Keating, K.A., and Cherry, S. 2004. Use and interpretation of logistic regression in habitat selection studies. *Journal of Wildlife Management* 68:774-789.

Krebs, J.R., and Davies N.B. 1996. *Behavioral Ecology: An Evolutionary Approach*. Fourth Edition. Sinauer Associates, Sunderland,MA.

Lankau, R., Søgaard Jørgensen, P., Harris. D.J., and Sih, A. 2011. Incorporating evolutionary principles into environmental management and policy. *Evolutionary Applications* 4:315–325.

Latif, Q. S., Heath, S. K., and Rotenberry, J. T. 2011. An 'ecological trap' for yellow warbler nest microhabitat selection. *Oikos* 120: no. doi: 10.1111/j.1600-0706.2010.18835.x

Latombe, G., Parrott, L., and Fortin, D. 2011. Levels of emergence in individual based models: Coping with scarcity of data and pattern redundancy. *Ecological Modelling* 222:1557-1568.

Levin, S.A. 1992. The problem of pattern and scale in ecology. *Ecology* 73:1943–67.

Lima, S. L., and Zollner, P.A. 1996. Towards a behavioral ecology of ecological landscapes. *Trends in Ecology and Evolution* 11:131–134.

Long, E.S., Diefenbach, D.R., Rosenberry, C.S., and Wallingford, B.D. 2008. Multiple proximate and ultimate causes of natal dispersal in white-tailed deer. *Behavioral Ecology* 19(6):1235–1242.

McLane, A.J., Semeniuk, C.A.D., McDermid, G.J., and Marceau, D.J. 2011. The role of agent-based models in wildlife ecology and management. *Ecological Modelling* 22:1544-1556.

Marceau, D.J. 2008. What can be learned from multi-agent systems? In: Gimblett, R. (Ed.), *Monitoring, Simulation and Management of Visitor Landscapes.* University of Arizona Press, pp. 411–424.

Marmion, M., Luoto, M., Heikkinen, R.K., and Thuiller, W. 2009. The performance of state-of-the-art modelling techniques depends on geographical distribution of species. *Ecological Modelling* 220:3512-3520.

May, R., van Dijk, J., Landa, A., and Andersen, R. 2010. Spatio-temporal ranging behaviour and its relevance to foraging strategies in wide-ranging wolverines. *Ecological Modelling* 221:936-943.

Mayor, S.J., Schneider, D.C., Schaefer, J.A., and Mahoney, S.P. 2009 Habitat selection at multiple scales. *Ecoscience* 16:238–247.

Morales, J. M., and Ellner, S.P. 2002. Scaling up movements in heterogeneous landscapes: the importance of behavior. *Ecology* 83:2240–2247.

Mosser, A., Fryxell, J. M., Eberly, L., and Packer, C. 2009. Serengeti real estate: density vs. fitness-based indicators of lion habitat quality. *Ecology Letters* 12:1050–1060

Musiani, M., Morshed Anwar, S., McDermid, G.J., Hebblewhite, M., and Marceau, D.J. 2010. How humans shape wolf behavior in Banff and Kootenay National Parks, Canada. *Ecological Modelling* 221(19): 2374–2387.

Nathan, R., Getz, W.M., Revilla, E., Holyoak, M., Kadmon, R., Saltz, D., and Smouse, P.E., 2008. A movement ecology paradigm for unifying organismal movement research. *Proceedings of the National Academy of Sciences* 105:19052–19059.

Parker, D.C., Manson, S.M., Janssen, M.A., Hoffmann, M.J., and Deadman, P. 2003. Multiagent systems for the simulation of land-use and land-cover change: a review. *Annals of the Association of American Geographers* 93(2):314-337.

Patterson, T. A., Thomas, L., Wilcox, C., Ovaskainen, O., and Matthiopoulos, J. 2008. State-space models of individual animal movement. *Trends in Ecology and Evolution* 23:87–94.

Pérot, A., and Villard, M-A. 2009. Putting density back into the habitat-quality equation: case study of an open-nesting forest bird. *Conservation Biology* 23:1550–1557.

Pettifor, R., Caldow, R., Rowcliffe, J., Goss-Custard, J., Black, J., Hodder, K., Houston, A., Lang, A., and Webb, J. 2000. Spatially explicit, individual-based, behavioral models of the annual cycle of two migratory goose populations. *Journal of Applied Ecology* 37:103–135.

Piou, C., Berger, U., and Grimm, V. 2009. Proposing an information criterion for individual-based models developed in a pattern-oriented modelling framework. *Ecological Modelling* 220:1957–1967

Railsback, S.F., and Harvey, B.C. 2002. Analysis of habitat-selection rules using an individual-based model. *Ecology* 83:1817–1830.

Rands, S.A., Pettifor, R.A., Rowcliffe, J.M., and Cowlishaw, G. 2004. State-dependent foraging rules for social animals in selfish herds. *Proceedings of the Royal Society London B* 271:2613–2620.

Revilla, E., and Wiegand, T. 2008. Individual movement behavior, matrix heterogeneity, an the dynamics of spatially-structure populations. *Proceedings of the National Academy of Sciences* 105:19120–19125.

Rossmanith, E., Grimm, V., Blaum, N., and Jeltsch, F. 2006. Behavioural flexibility in the mating system buffers population extinction: lessons from the lesser spotted woodpecker *Picoides minor*. *Journal of Animal Ecology* 75:540-548.

Russell, S.J., and Norvig, P. 1995. Intelligent Agents. In: Russell, S.J., and Norvig, P. (Eds.), *Artificial Intelligence; A Modern Approach.*, Prentice Hall, Englewood cliffs, New Jersey, UK. pp. 33-51.

Schtickzelle, N., Gwénaëlle M., and Baguette, M. 2006. Dispersal depression with habitat fragmentation in the bog fritillary butterfly. *Ecology* 87:1057–1065.

Stamps, J.A. 2009. Habitat Selection. In: Levin, S.A. (Ed.), *The Princeton Guide to Ecology*. Princeton University Press, New Jersey, USA. pp. 38-44.

Stamps, J.A., and Groothuis, T.G.G. 2010. The development of animal personality: relevance, concepts and perspectives, *Biological Reviews* 85:301–325.

Tang, W. 2008. Simulating complex adaptive geographic systems: a geographically-aware intelligent agent approach. *Cartography and Geographic Information Science* 35:239–263.

Tang, W., and Bennett, D.A. 2010. Agent-based modeling of animal movement: a review. *Geography Compass* 4(7):682–700.

Topping, C.J., Dalkvist, T., Forbes, V.E., Grimm, V., and Silby, R.M. 2009. The potential for the use of agent-based models in ecotoxicology. In Devillers, J. (Ed.), *Ecotoxicology*

Modeling, Emerging Topics in Ecotoxicology: Principles, Approaches and Perspectives 2. Springer, New York. pp. 205-235

Tyre, A.J., Kerr, D.G,. Tenhumberg, B., and Bull, C.M. 2007. Identifying mechanistic models of spatial behaviour using pattern-based modelling: an example from lizard home ranges. *Ecological Modelling* 208:307–316.

Valcu, M., and Kempenaers, B. 2010. Spatial autocorrelation: an overlooked concept in behavioral ecology. *Behavioral Ecology* 21(5):902-905.

Willems, E.P., Barton, R.A., and Hill, R.A. 2009. Remotely sensed productivity, regional home range selection, and local range use by an omnivorous primate. *Behavioral Ecology* 20(5):985-992.

Yoder, J.M., Marschall, E.A., and Swanson, D.A. 2004. The cost of dispersal: predation as a function of movement and site familiarity in ruffed grouse. *Behavioral Ecology* 15:469-476.

Zollner, P.A., and Lima, S.L. 2005. Behavioral trade-offs when dispersing across a patchy landscape. *Oikos.* 108:219-230.

2

Evolution of Ecosystem Services in a Mediterranean Cultural Landscape: Doñana Case Study, Spain (1956-2006)

Erik Gómez-Baggethun[1,2], Berta Martín-López[2], Pedro L. Lomas[2], Pedro Zorrilla[2] and Carlos Montes[2]
[1]*Institute of Environmental Science and Technology, Universitat Autònoma de Barcelona*
[2]*Social-Ecological Systems Laboratory, Department of Ecology, Universidad Autónoma de Madrid*
Spain

1. Introduction

The conceptualization of ecosystems as natural capital that provides services to society, to some extent emerges as a strategic or pragmatic attempt to put in value the role that nature plays in human well-being. Several classifications of the benefits that society receives from ecosystems have been developed in the scientific literature, both in terms of services and in terms of functions or using both concepts with different connotations (King, 1966; Daily, 1997; Costanza et al., 1997; De Groot et al., 2002; Douguet & O'Connor, 2003; Naveh, 2004). These classifications have also been used at international projects such as the CRITINC project (Van der Perk & De Groot, 2000), the Millennium Ecosystem Assessment (MA, 2003), and the initiative The Economics of Ecosystems and Biodiversity (Kumar (ed.), 2010).

Assessing ecosystem services involves, for analytical purposes, the translation of complex and interlinked ecological structures and processes into a limited number of ecosystem functions that in turn provide diverse services for humans at different scales (De Groot, 1992, 2006; De Groot et al., 2002). The main difference between ecosystem functions and ecosystem services is related to the direct enjoyment, consumption or use by humans. Sometimes ecosystems generate ecosystem functions that are neither demanded nor valuated by humans (e.g. remote inhabited and unexploited ecosystems) and thus do not strictly involve the supply of ecosystem services except a few global scale services such as carbon sequestration or biodiversity conservation. In this context, ecosystem functions refer to potential services, or to the ecosystems *capacity* to provide services, while the concept of ecosystem services entails that these have current *value* for society (Gómez-Baggethun & de Groot, 2010).

This research draws on the conceptual framework of the Millennium Ecosystem Assessment (MA), which distinguishes four different categories of ecosystem services: life-support, regulating, cultural and provisioning services (MA, 2003). Nevertheless, the delineation between the categories of *regulating* and *life-support* services is often ambiguous (MA, 2003;

Martín-López et al., 2009). Furthermore, as has been argued by Hein et al. (2006), the consideration of life-support services as a separate category might lead to double-counting problems with other categories of services (Boyd & Banzaff, 2007; Fisher et al., 2009). For this reason, large-scale ecological processes such as primary production, water cycle or biogeochemical cycles have been conceptualized in our study as core ecosystem processes, whose performance is considered a necessary precondition for the generation of the other categories of services that are relevant from a human perspective, namely regulating, cultural and provisioning services (Figure 1).
Therefore, *life-support* services, included in the MA's conceptual framework, have not been directly addressed in this study. As shown in Figure 1, almost every form of social wealth, as well as the different aspects of human well-being is in some way nurtured by, or dependent on, the ecosystems tangible and intangible services. This is the basis of the approach we use in this paper, which emphasizes the role of ecosystems in human well-being, not only when subject to exploitation in order to obtain provisioning services, but also when they are preserved, since regulation functions are better maintained.

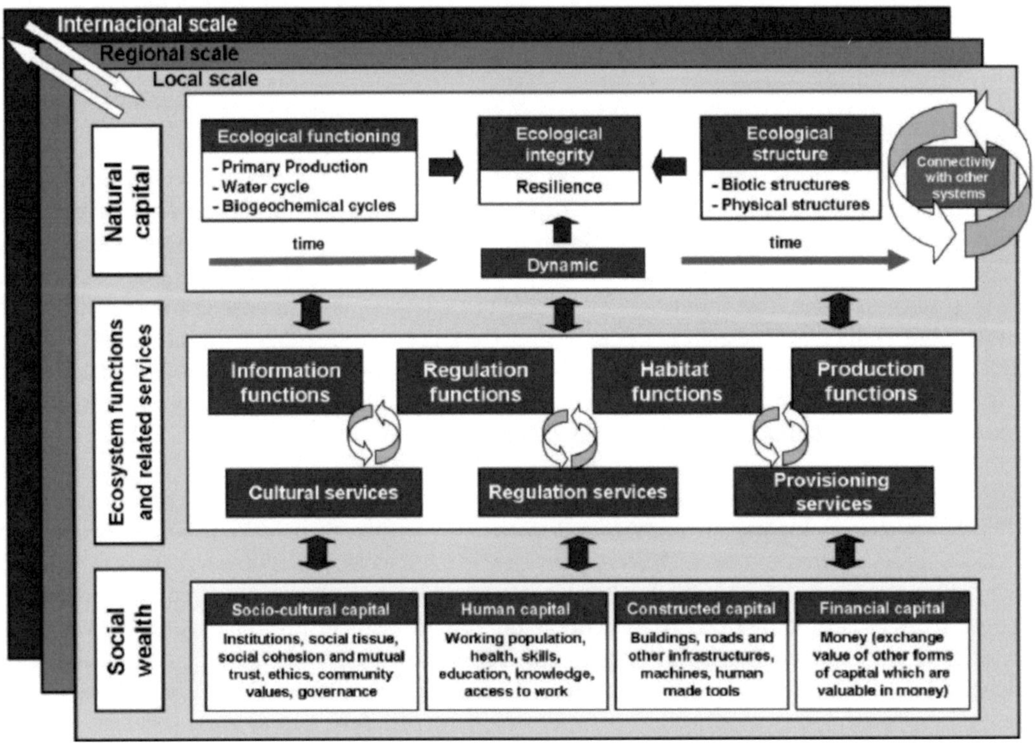

Fig. 1. Ecosystems can be conceptualised from an anthropocentric perspective as a form of natural capital, performing several functions that in turn generate multiple goods and services that are enjoyed by stakeholders at different scales. Ecosystem services nurture every form of social wealth, conceptualised in the figure as different forms of capital.

The main goal of this study is to explore state, trends, and trade-offs in the evolution of the ecosystem services flow provided by the ecosystems of Doñana (SW Spain), which is considered the most relevant wetland in Spain. With this aim, the current state of Doñana's ecosystem services as well as their trends during the last fifty years (1956-2006) have been approached. The studied period encompass a critical phase due to the intensity of the transformations undergone in this area, as it covers the period within which Doñana transits from subsistence to a market-oriented economy, involving deep institutional changes in the way ecosystem services are produced and distributed (Gómez-Baggethun and Kelmen, 2008; Gómez-Baggethun, 2010; Martín-López et al., 2011).

2. The Mediterranean context

The Mediterranean basin is considered as a transitional climatic area between the subtropical desert belt and the more humid northern domain. Ecosystems at the Mediterranean basin have co-evolved through millennia with different cultures generating Mediterranean landscapes (Blondel, 2006). Resource use and transformation is so ancient in this region that Naveh & Lieberman (1993) suggested there are no strictly natural landscapes in the Mediterranean basin any more, arguing that it is more accurate to talk of cultural landscapes. In fact, today's Mediterranean landscapes have been shaped by more than eight millennia of an agro-silvo-pastoral way of life (Grove & Rackham, 2003; Butzer, 2005). This way of life has progressively modelled multi-functional landscapes, often based on agro-silvo-pastoral systems of polyculture. The fact that biodiversity hotspots have been able to emerge within highly humanized landscapes providing diverse ecosystem services, witnesses a successful long term nature-society co-evolutionary process in the Mediterranean basin.

Nevertheless, during the last decades, Mediterranean cultural landscapes have been subject to increasing pressures, and thus are being transformed at unprecedented rates of change (Pinto-Correia & Vos, 2002). Some of these changes are manifested in terms of the homogenization of landscape use and loss of ecosystem services (Brandt & Vejre, 2002). Cultural homogenization dynamics inherent to the globalization process are being accompanied by other large scale drivers such as industrialization or the introduction of economies of scale into productive land use functions, resulting in the transformation of multi-functional cultural landscapes into more simplified spatial patterns dominated by mono-functional land use (Brandt & Vejre, 2002) and thus, in impoverished flows of ecosystem services. It seems to be some consensus about the idea that the current process of global change is carrying landscape homogenization and an increasing conversion of natural and seminatural ecosystems. Nevertheless, this general idea should be tested in concrete case studies to better analyze their causes and impacts, as we do in this study.

3. The Doñana case study

We conceptualized Doñana region as a social-ecological system due to the tight cultural and economic links between its natural area and the human population of 16 municipalities of three different provinces of Andalusia: Seville, Huelva and Cadiz. Doñana is considered to be one of the most emblematic wetlands in Western Europe and encompasses two important protected areas, the Doñana National Park and the Doñana Natural Park (Figure 2), which are both highly appreciated for their ecological and cultural values. Doñana is a unique natural area in many aspects: it is a major stopover point in the migration route of birds moving between Europe and Africa, it provides habitat for one of the most endangered mammals in the world –the Iberian lynx (*Lynx pardinus*)-, as well as for many endemic, threatened and ecologically interesting species; it constitutes perhaps the most outstanding and better studied wetland in western Europe (Fernández Delgado, 2005). As shown in Figure 2, the ecological limits of Doñana correspond with the fluvial-littoral ecosystem of Doñana (2,205 km2), a wide system of marshes, dunes and beaches, associated with the coastal dynamic of the Guadalquivir River's mouth (Montes et al. 1998). From a hierarchical analysis of their ecosystems (Klijn & Haes, 1994), the Greater fluvial-littoral ecosystem of Doñana encompasses four different types of ecosystem units: marshes (1,591 km^2), aeolian mantles (505 km^2), coastal system (39 km^2) and estuary (69 km^2).

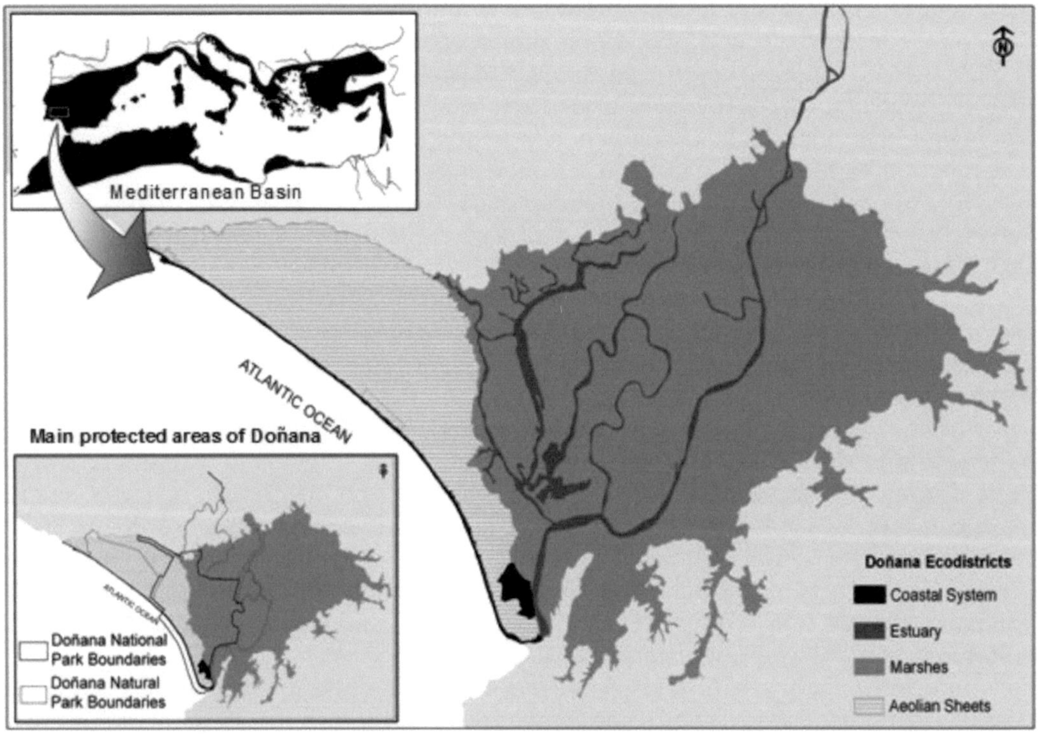

Fig. 2. The fluvial-littoral ecosystem of Doñana is located in the western part of the Mediterranean basin. It embeds four different ecosystems at scale of ecodistrict: coastal system, estuary, marshes and aeolian sheets as well as two important protected areas: Doñana National Park and Doñana Natural Park (unified in 2005 as Doñana Natural Area).

As stated by Rodríguez Merino & Cobo García (2002), Doñana has historically been subject to a wide range of traditional economic uses coupled to local ecosystem's dynamics (Ojeda-Rivera, 1987; Gómez-Baggethun, 2010). The shifting mosaic (*sensu* Forman, 1995), based on multiple uses of the territory has been the dominant landscape management model in Doñana until a few decades ago (Ojeda-Rivera, 1987). Due to its isolation and the marginal character of its land (nutrient poor sandy soils and braquish marshlands), large-scale territorial transformation arrived late to this area (González-Arteaga, 1993). While in most European countries large wetlands had been dried out during the 18th and 19th centuries in order to control malaria and increase land productivity, all trials of reclamation of Doñana's marshes until the 20th century had failed due to lack of technology and capital investment

(González-Arteaga, 1993). During the first decades of the 20th century, Doñana was therefore an almost unique case of wetland conservation in the European context. Furthermore, Doñana was at that time a feeble populated and almost isolated area, which actually had no access road, with a subsistence-oriented economy based on multiple landuses (Ojeda-Rivera, 1990; Villa-Díez et al., 2000). This situation started to change in 1929, with a transformation process that involved the progressive deployment of four, often conflicting, different management policies: agriculture, forestry, tourism and conservation (Montes, 2000).

Between 1929 and 1956, private companies started to drain parts of the marsh in order to cultivate rice (González-Arteaga, 1993). The transformation process was accelerated, through State reclamation projects during the 1956-1978 period, when the upper and part of the lower marsh was drained for further agricultural purposes. In the same period, the State implemented an extensive forest plan to replant the aeolian mantles with eucalyptus (*Eucalyptus* spp.), destroying more than half of the cork tree forest, and a major project to irrigate crop with groundwater was initiated, affecting the aeolian mantles' water regulating functions (Custodio, 1995; García-Novo & Marín-Cabrera, 2005). Development projects in the coast were deployed from 1969, when the beaches of the area were declared of national interest for tourism, resulting in the major urbanization of the coastal area of Matalascañas. Finally, during the 20th century, the Guadalquivir River branches were progressively channeled in order to shorten the navigation distance to Sevilla through the estuary (Menanteau, 1984; González Arteaga, 2005). The period considered in this study, 1956-2006, thus coincides with a transformation process that often involved the simplification of ecosystems by command and control management strategies aimed to increase the productivity by the enhancement of intensive mono-functional land uses.

As a response to this fast transformation process, at the end of the 1960's conservationist policies promoted by European institutions and national and international conservationists were deployed in Doñana. Since the declaration of Doñana as National Park in 1969, protected areas in Doñana have been extended up to now through the declaration of new protection categories and through the enlargement of the existing protected areas. The aim has been to preserve remaining habitats of flagship species in a context of powerful development interests (Figure 3).

Nevertheless, the arrival of strict conservationist policies to Doñana also entailed the prohibition of many socio-economic activities within the protected areas, except those related to ecotourism and a few traditional uses, affecting the flow of provisioning services and the stakeholders whose livelihoods were related to ecosystem production functions. As a consequence, during the last few decades Doñana has been subject to increasing subsidies in order to attenuate social conflicts emerging in relation to conservationist restrictions. Following Ojeda-Rivera (1993), the permanent flow of subsidies, often foreign to the existing local socio-economic tissue, has derived in the establishment of a subsidized culture in Doñana that discourages initiatives for endogenous development. The implementation of strict conservation strategies in Doñana has therefore had different effects. On the one hand, conservation policies have managed to slow down the ecosystem transformation process, for instance achieving to stop the urbanization of the coast, the further reclamation of remaining natural marshes, and the development of linear infrastructures with high impact on habitat fragmentation. On the other hand, by putting strict constraints to most socioeconomic activities, conservation policies (paradoxically like development policies) have also contributed to the erosion of the system of multiple uses in multifunctional landscapes.

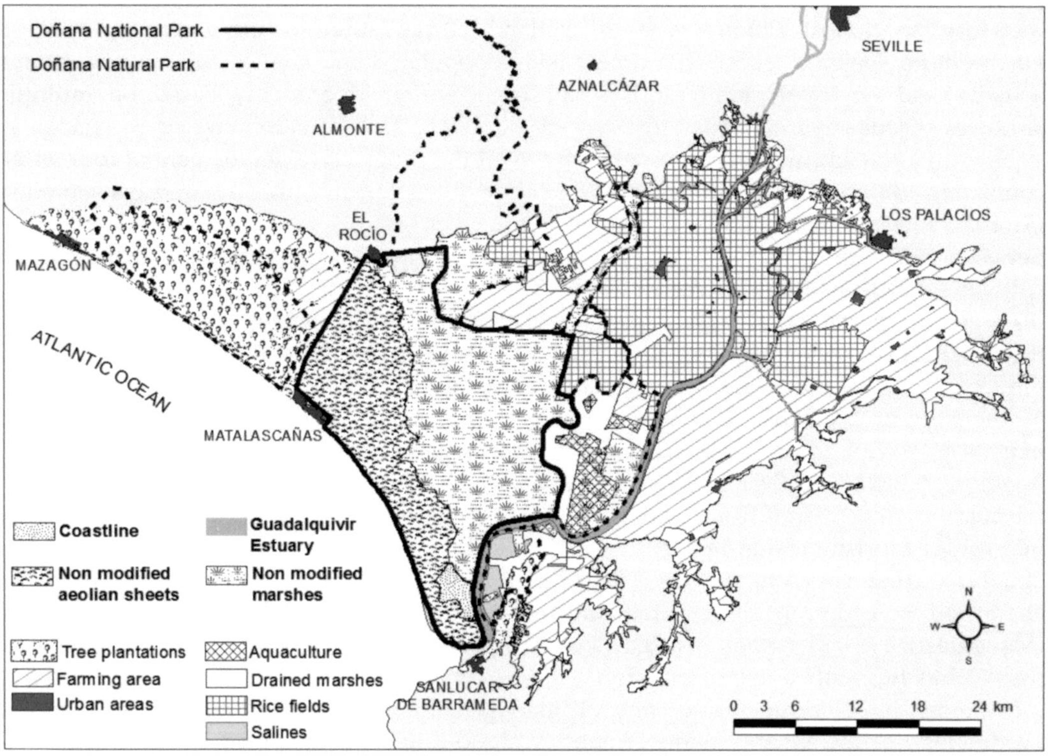

Fig. 3. Main land uses within the Greater fluvial-littoral ecosystem of Doñana. Almost every surface surrounding the protected areas has been transformed.

To sum up, four uncoordinated, and often competing policies (agriculture, forestry, tourism and conservation) were deployed during the 20th century. Conversion of natural ecosystems and subsequent effects on the flow of ecosystem services happened in the absence of an integrated territorial planning in Doñana. In this context Doñana has been portrayed as a clear example of the conservation versus development paradigm in territorial planning where green fortress-protected areas emerge in a matrix of degraded territories devoted to economic development (Gómez-Baggethun et al., 2010; Martín-López et al., 2011).

4. Methods

4.1 Characterization of drivers of change

An essential step of ecosystem services assessments is to characterize and measure through proxy data or indicators the main drivers of change operating in the area. Our characterization of drivers of change draws on quantitative data from official national (National Statistics Institute) and regional (Andalusian Statistics Institute, SIMA) statistics offices, as well as from GIS analysis of land cover changes in Doñana during the period 1956-2006 using aerial photographs and Landsat TM Imagery (Zorrilla et al., forthcoming).

Four drivers of change were characterized using either quantitative data or proxy indicators: population growth, changes in labor structure, conservation policies, and development policies. Each driver was quantified using one or more indicators as proxy measures. More specifically, population growth was measured as variation in the number of inhabitants, changes in labor structure was measured through changes in the relative importance of the agricultural, industrial and service sectors, conservation policies were approached through the variation in the total protected area as well as in the number of protected areas, and the importance of development policies was approached through the increase in the length of lineal infrastructures as well as through the increase in the total area covered with spatial infrastructures (mainly urban areas). When data were not available for the whole period, a representative period was selected.

4.2 Assessment of ecosystem services state and trend

State and trend in regulating, cultural and provisioning services were assessed separately for the four ecodistricts of Doñana. Relevant ecosystem services were characterized following previous research in the area (Gómez-Baggethun, 2010; Martín-López et al., 2010), and supported by an in-depth literature review, scanning of administrative documents, and fieldwork interviews with local resource users, managers, scientists and other key informants conducted during 2006.

As the assessment of changes in ecosystem services at the scale of ecodistrict required abundant data and expertise criteria, the assessment of the services state and trend was entrusted to a scientific expert panel. Ecosystem services of each ecodistrict were evaluated by a panel of 10 scientists, including researchers from five different Spanish universities as well as staff from the Spanish National Research Council (CSIC), which has carried out research in the Biological Reserve of Doñana since 1968. Every member of the panel had no less than 8 years of research experience in Doñana. This multidisciplinary panel, which included specialists in Biology, Ecology, Hydrology, Limnology, Geomorphology, Environmental Sciences, Economy and Social Sciences, assessed current state of Doñana's ecosystem services in a qualitative Likert scale: very degraded (0), degraded (1), adequate (2), good (3), and very good (4).

Next, trends in the ecosystem services were assessed in order to study their evolution in the period 1956-2006. As in the case of the assessment of state, trends were analyzed using a five step Likert scale ranging from strongly deteriorated to strongly improved performance as follows: strongly deteriorated (0), deteriorated (1), stable (2), improved (3) and strongly improved (4). Results were analyzed using statistical methods.

Finally, in both cases, we used non-parametric statistics (Kruskal-Wallis test) to determine differences of ecosystem services among ecodistricts. Additionally, we used Mann-Whitney tests to determine differences between short scale (locally orientated) and large scale (orientated at the national and the international scales) supply of provisioning and cultural services. This scale differentiation was done in order to check if the services flow was mainly orientated to the Doñana community or if it was rather orientated to satisfy the demand from stakeholders at broader scales.

5. Results

5.1 Drivers of change

Tendencies in the four drivers of change considered in this study, i.e. population growth, economic transition towards the services sector, deployment of conservationist policies, and infrastructure development, are shown in Figure 4.

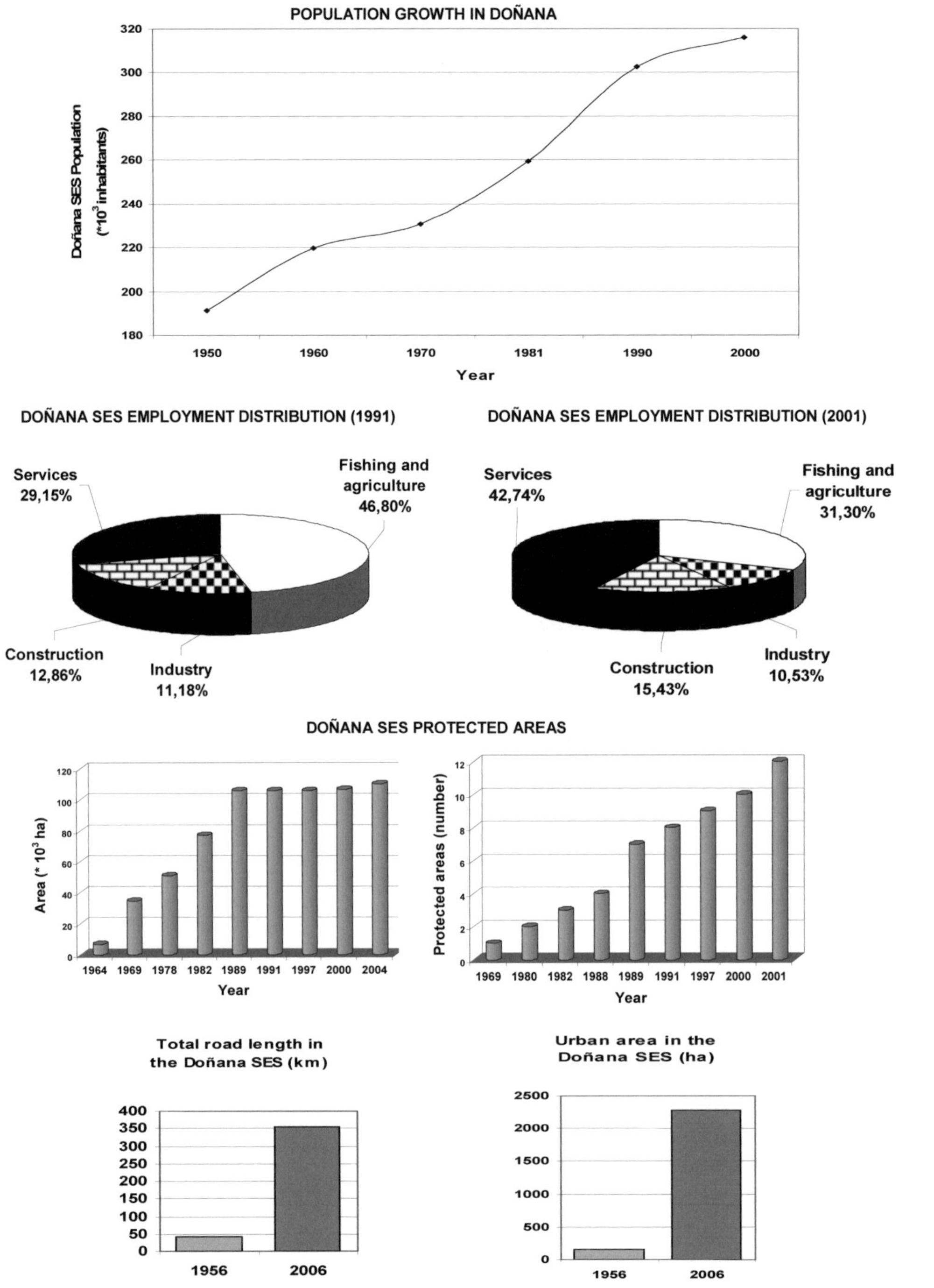

Fig. 4. Population growth, economic transition towards services sector, conservation policies and infrastructure development are among the most powerful drivers acting on the transformations in land use and ecosystem services in Doñana during the studied period. Source: Own development from data of the Andalusian Statistics Institute.

First, population trends show a steady growing trend throughout the studied period, growing from less than 200,000 inhabitants by the 1950s to more than 300,000 inhabitants in the 2000s.

Second, the data show a fast growth of the secondary (industrial plus housing) and tertiary (services) sectors at the expense of the primary sector (agriculture and fishery). Only in the 1991-2001 period, the relative importance of the primary sector in the economy of Doñana (proxied through share of employment) diminished from 47% to 31% of the total number of employments. The secondary sector increased moderately from 24% to 26% of total employment, whereas the tertiary sector increased from 29% to 43% of total employment, showing a marked tertiarization of Doñana's economy throughout the studied period.

Third, our data show the importance of nature conservation as a key driver of change throughout the studied period. Total protected area increased from about 6800 ha in 1964 to more than 115,000 ha in 2004, whereas the number of protected areas increased from zero to 12 throughout the studied period.

Finally, our data suggest a great importance of development policies as a critical driver of territorial change throughout the studied period. The variation in the length of lineal infrastructures shows an increase from less that 50 km in 1956 to more than 350 km in 2006, whereas the total surface of urban areas increased from less that 200 ha in 1956 to about 2300 ha in 2006.

5.2 Ecosystem services state and trend

At the ecodistrict scale, 23 relevant services were found to be provided by the marshes, 24 by the aeolian mantles, 16 by the coastal system, and 22 by the estuary (Table 1). The assessment of the ecosystem services state and trend was conducted independently for each ecodistrict. The assessment responds therefore to a general picture of the ecosystem services of each ecodistrict, irrespective of which part of them was inside the protected areas and which was not.

Marsh		
Regulating services	*Cultural services*	*Provisioning services*
Sedimentary balance	Recreation and ecotourism	Food and fiber crops cosmetic plants
Nutrient regulation	Landscape beauty and aesthetical values	Livestock
Surface / ground water flow regulation	Cultural heritage and sense of place	Gathering
Flood buffering/	Didactic, educative and interpretative functions	Fishing
Climate control	Local ecological knowledge	Aquiculture
Breeding and refugee of migratory species	Scientific research	Medicinal / aromatic plants
Detoxification and pollution processing		Salt works
Maintenance of the saline equilibrium		Land for construction
		Employment

Aeolian sheets		
Regulating services	*Cultural services*	*Provisioning services*
Erosion control	Recreation and ecotourism	Fresh water
Peat formation / maintenance	Landscape beauty	Food crops and plantations
Maintenance of dune dynamic	Cultural heritage and sense of place	Livestock
Maintenance of wetlands	Didactic, educative and interpretative functions	Hunting
Surface / ground water flow and salt regulation	Scientific research	Gathering
Detoxification	Local ecological knowledge	Materials: wood, cork, resin
Nutrient regulation		Fuel: wood, coal, pines
Pollination		Honey and beekeeping
Soil formation		Land for construction
		Employment
Coastal system		
Regulating services	*Cultural services*	*Provisioning services*
Erosion control and sediment retention	Recreation and beach tourism	Seafood
Coastal stabilization	Landscape beauty and aesthetical values	Fishing
Storm and wave buffering	Cultural heritage and sense of place	Land for construction
Climate control	Didactic, educative and interpretative functions	Employment
Detoxification and pollution processing	Scientific research	
Maintenance of habitats and food webs	Local ecological knowledge	
Estuary		
Regulating services	*Cultural services*	*Provisioning services*
Erosion control	Recreation and ecotourism	Seafood
Coastal dynamic regulation: sediment retention / movement	Landscape beauty and aesthetical values	Fishing
Surface / ground water flow and salt regulation	Cultural heritage and sense of place	Aquiculture
Flood buffering	Didactic, educative and interpretative functions	Hunting
Regulating services	*Cultural services*	*Provisioning services*
Detoxification and pollution processing	Scientific research	Salt works
Nursery	Local ecological knowledge	Employment
Maintenance of habitats and food webs		
Maintenance of the saline equilibrium		

Table 1. Main ecosystem services provided by the ecosystems of Doñana.

Results showed the category of regulating services to be the most affected one, as mean values of state show some degree of degradation in all the four ecodistricts (Figure 5). Kruskal-Wallis test showed significant differences for the state of regulating services among ecodistricts (χ^2 = 8.01, p = 0.04). State results of regulating services showed the estuary to be the most degraded ecodistrict, while those supplied by the coastal systems were adequate on average according to the scientific panel. While there was no significant difference among ecodistricts regarding trends in regulating services (χ^2 = 6.32, p = 0.09), results also show generalized, though moderate, deterioration of regulating services except in the case of the coastal system, where trends suggest stability in performance. Deterioration is considerable in the case of the aeolian sheets and moderate in the marshes and the estuary (Figure 5).

The category of cultural services showed the most positive results in both state and trend. Mean state values are adequate to good in all the ecodistricts, without significant differences between them (χ^2 = 5.17, p = 0.16). In contrast, there are differences among ecodistricts for the trend variable (χ^2 = 6.80, p = 0.07). The estuary is the ecodistrict which has suffered the most significant deterioration of cultural services. It should be noted, however, that even though cultural services are the best maintained in Doñana, there are significant differences for the state (U = 18.0, p = 0.014) and trend (U = 16.0, p = 0.04) between those closely related to local culture and those whose use value is related to stakeholders at national and international scales (Figure 6). The cultural services related to the traditional ecological knowledge and sense of place are the most degraded, while services related to scientific research and tourism seem to have improved during the last decades as they have been permitted and promoted by conservation policies.

Concerning the provisioning services, results showed no significant differences among ecodistricts for state (χ^2 = 2.51, p = 0.47) and trend (χ^2 = 5.25, p = 0.15). Results of the state variable showed adequate levels of performance in all the ecodistricts, except in the case of the aeolian mantles, where mean state values suggest ecosystem services to be slightly degraded (Figure 5). Results in provisioning services trends were lower (more degraded) on average, as some deterioration is found in all the ecodistricts except the coastal system, where the trend seems to be of stability on average. Similarly to what our results showed for cultural services, within provisioning services we found significant differences for state (U = 18.5, p = 0.013) and trend (U = 15.5, p = 0.047) when provisioning services related to local consumption and those which are primarily demanded by stakeholders at broader scales were compared (Figure 7). In accordance with what could be expected, local use of provisioning services has suffered important deterioration, as opposed to the provisioning services aimed at stakeholders related to the national and the international market, such as cash crops, which have improved during the analyzed period.

6. Discussion

6.1 Trade-offs within the flow of ecosystem services: Changing the scale

Significant qualitative changes were identified in the flows of ecosystem services provided by the ecosystems of Doñana during the period 1956-2006. In order to find general trends to characterise these changes, the scale at which the supply of ecosystem services is fostered, and the scale at which services are being demanded and used, seems to be one of the most relevant aspects in order to analyze the qualitative shift undergone by the ecosystem services flow (see e.g. Martín-López et al., 2010). As stated by several authors (MA, 2003, Hein et al., 2006;

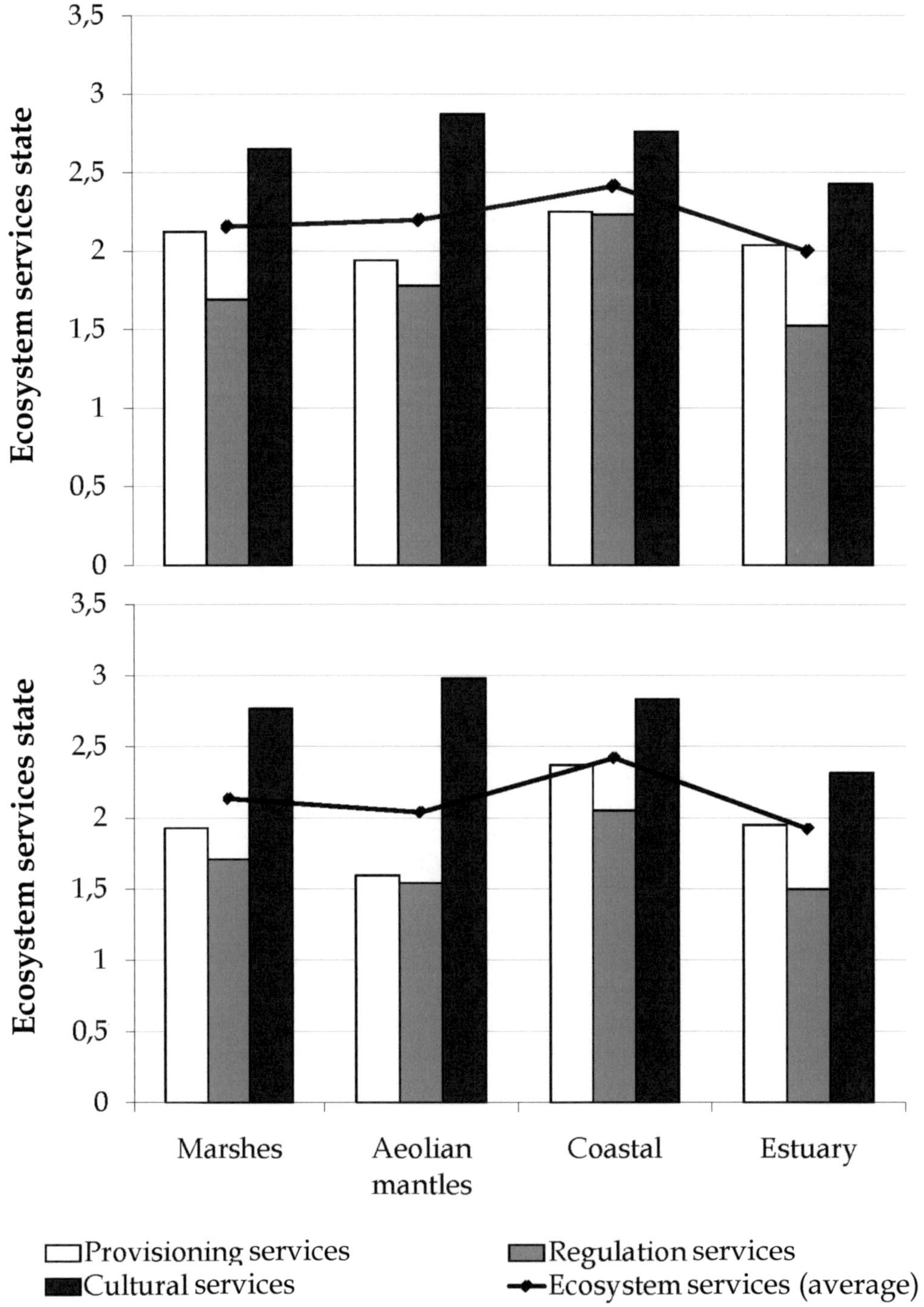

Fig. 5. Average values of the current state and trend (1956-2006 period) of the ecosystem services provided by the ecodistricts of Doñana. Cultural services are the category with best levels of performance, while regulating services appear to be the most degraded.

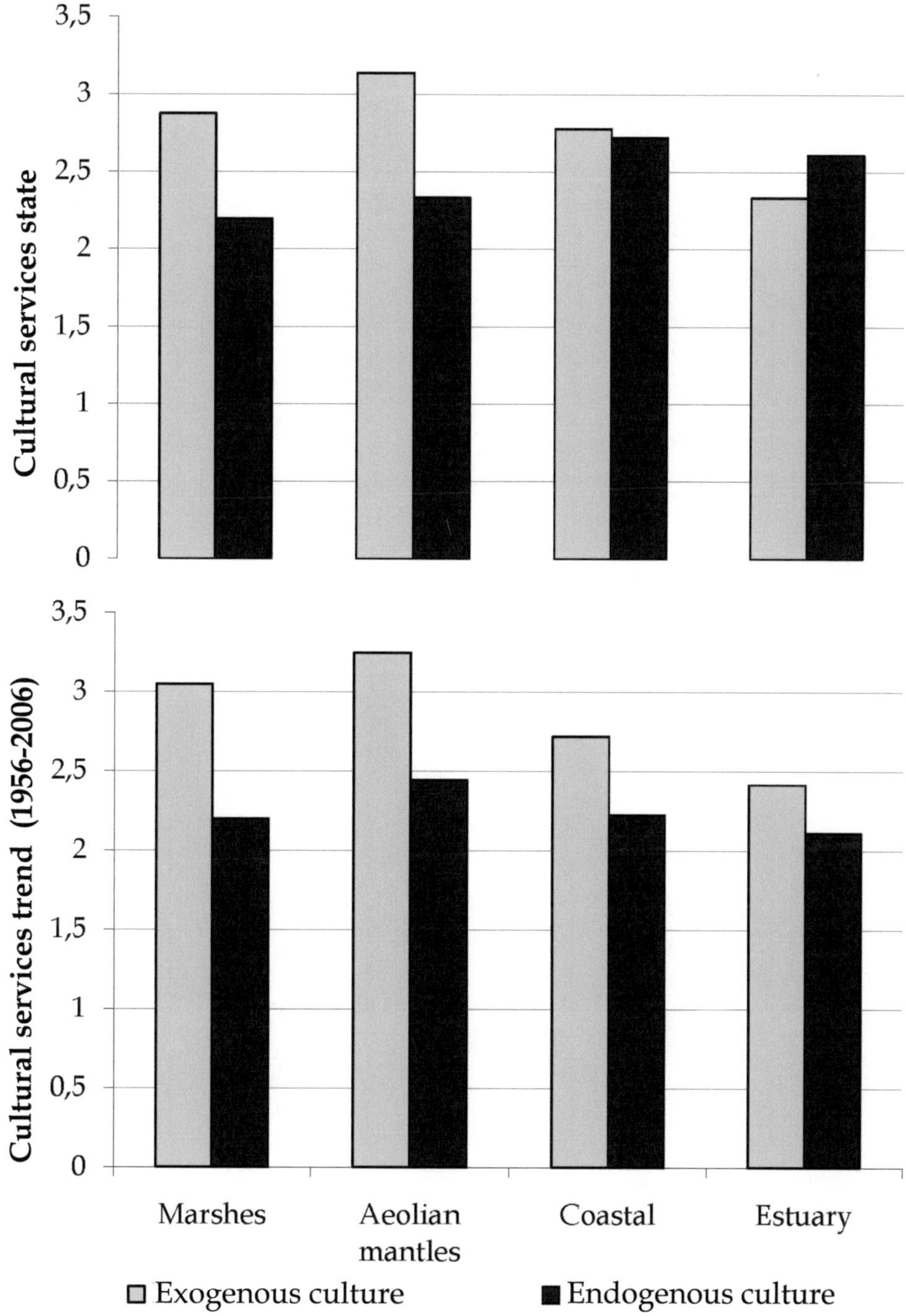

Fig. 6. Cultural service flows from Doñana's ecosystems are experiencing a delocalization process. Cultural services flows, primarily oriented to the local inhabitants at the beginning of the study period are becoming progressively commodified and oriented towards (sold to) stakeholders at national and international scales.

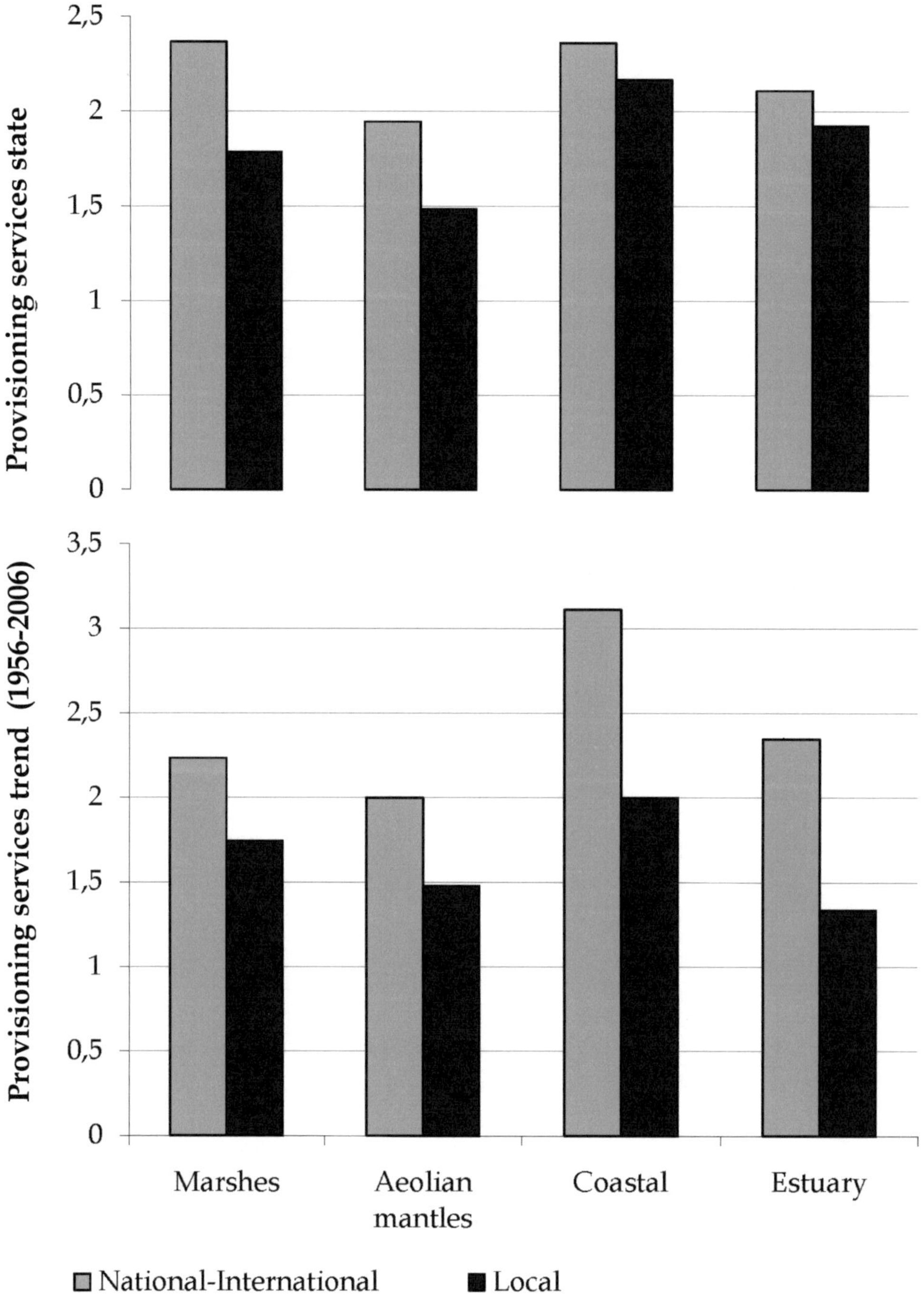

Fig. 7. Provisioning service flows from Doñana's ecosystems are experiencing a delocalization process. Provisioning services flows, primarily oriented to the local inhabitants at the beginning of the study period are becoming progressively commodified and oriented towards (sold to) stakeholders at national and international scales.

Martín-López et al., 2011), there is a need to examine the different scales at which ecosystem services are provided and enjoyed, and how the supply of ecosystem services affects the interests of stakeholders from the local to the global scale.

In our study, this aspect becomes especially clear in the case of cultural and provisioning services. While general results of the cultural services valuation show adequate states and even improving trends, it is interesting to see how these services have evolved when the subcategories are analysed separately (Figure 6). In fact, when we zoom to cultural services subcategories, we can identify significant trade-offs. These trade-offs reflect a progressive transition from a management strategy primarily aimed at obtaining services to be used by the community of local stakeholders, towards a management strategy with special focus on services demanded by stakeholders at broader scales, such as national or international consumers, tourists, scientists or conservationists. Thus, while cultural identity, sense of place, traditional ecological and other cultural services orientated to the Doñana local community are being lost or degraded (see also Gómez-Baggethun et al., 2010), those cultural services orientated to actors outside the Doñana community such as science, research and environmental educational services, recreation and ecotourism, have improved.

As explained in section 5, something similar happens with the provisioning services (Figure 7). Hunting, gathering, forestry uses and other ecosystem services that historically were directly enjoyed by local communities show decreasing trends. This can be related, on one hand, to the decline of the community-based economy due to the increasing integration of Doñana in the national and international markets as a part as the current globalization process (Martín-López et al., 2010). On the other hand, conservationist related constrictions have also played a role in the decline of certain provisioning services, as they have not only affected industrial and intensive uses, but also small scale uses such as wood collection, hunting, gathering and other locally oriented services. In contrast, provisioning services demanded by stakeholders at broader scales such as rice crops and other forms of intensive agriculture and aquiculture are being enhanced (Figure 7). The results obtained by the expert panel concerning this issue, were consistent with the declarations of local resource users in fieldwork interviews. To sum up, the evolution of the ecosystem services flow generated by the ecosystems of Doñana during the half past century reflects the transition from a *community-based* economy, based on the needs of local stakeholders, towards a *market-based* economy, primarily orientated to satisfy the demands from stakeholders at broader scales.

Our results are consistent in these respect with findings from recent ecosystem services research conducted in Doñana using alternative data sets (biophysical accounts, monetary valuation, and other quantitative measures) (see e.g. Gómez-Baggethun et al., 2010; Martín López et al., 2010, 2011).

6.2 Considering the results within the weak vs. strong sustainability debate

In the context of the transition towards a market economy in Doñana stated above, marketed and high added value ecosystem services are increasing their performance at the expense of those that are not, which are being lost or have declined. Regulating services, usually non-marketed, show a pattern of generalized deterioration, as outside the protected areas cultural multi-functional landscapes have been widely converted into intensive exploitations. On the other hand, results of the assessment show similar patterns in the case of the cultural and the provisioning services. The direct enjoyment of the ecosystem services of Doñana seems to move from the local community towards stakeholders at broader scales. We can see that the Doñana community seems to be obtaining increasing income from their local ecosystems, while the direct enjoyment of the services provided by local ecosystems is decreasing. In other words, as Doñana integrates national and international markets, the benefits that the local population obtains from their ecosystems seem to be moving from the enjoyment of their *use value* towards obtention of *exchange value* from them (see also Gómez-Baggethun & Kelemen, 2008). It should be noted that this does not necessarily entail a decrease in the use value finally obtained (at least indirectly) by the community stakeholders from their local ecosystems. The shift lies in the fact that the Doñana community does not manage anymore its ecosystems with the aim of satisfying local needs directly from local ecosystem services. Rather, ecosystem services management strategies have become increasingly oriented towards obtaining increased income from commodified ecosystem services, that in turn is used by local beneficiaries to purchase (through the markets) goods and services provided by ecosystems worldwide. This transition in the exploitation model from a *community-based* to a *market-based* economy has implications in terms of social-ecological decoupling, increased ecological debt and increased ecological footprint. While the analysis of these implications is far beyond the scope of this paper (for a thorough analysis of this phenomenon at the scale of the Spanish economy see Carpintero, 2005 and Lomas et al., 2008), we argue that the transition stated above fosters consumption patterns in the Doñana community that are increasingly foreign to the opportunities and limitations related to local ecosystems.

This tendency has promoted the conversion of multi-functional landscapes providing diverse and often non marketed services, into mono-functional landscapes based on the maximization of one or few ecosystem services embodying high added value (pulp, rice, tourism). Put it differently, our results show a steady trend towards the commodification of ecosystem services in Doñana. On one hand, homogenization dynamics related to this conversion might have significant consequences in terms of resilience loss (Martín-López et al., 2010). On the other hand, the assumed economic rationale behind these conversions would be probably challenged if the so-called negative environmental externalities were taken into account. The transformation of multi-functional into mono-functional landscapes generates increased private benefits (Balmford et al., 2002; de Groot, 2006), whereas the often higher costs, in terms of pollution, biodiversity extinction, and natural capital depletion are externalised to society at large or to future generations, thus not being considered in conventional economic accounting (Kapp 1983).

The transition process mentioned above can also be relevant within the weak versus strong sustainability debate (Neumayer, 1999). A thorough analysis in terms of weak and strong sustainability, would require further function analyses in which ecosystem services were evaluated and quantified in physical terms (see Martín-López et al., 2010, 2011 for substantial advances in this direction). Nevertheless, it is a fact that Doñana has lost important extensions of its natural capital stock since 1956, as outside the protected areas, ecosystems and cultural landscapes have been turned into simplified cultivated capital (i.e. monocultures of rice and eucalyptus plantations) and to a smaller extent also into constructed capital (industrial and urban areas) (González-Arteaga, 1993, Zorrilla et al., forthcoming). Furthermore, the picture of generalized deterioration of the regulating services reflected by the assessment of the expert panel shows how natural capital functions are being degraded. We can therefore argue that physical structures and processes of the

Doñana's natural capital are being degraded, while the monetary income obtained from Doñana's natural capital seems to be increasing. In this context, we could argue that Doñana is moving towards a weak sustainability path as its natural capital is replaced by financial and human-made capital.

7. Conclusions

During the last half century, the Doñana region has been subject to a process of deep structural transformation, in which the rates of change have been significantly accelerated. Demographic, socioeconomic, political and cultural drivers have played an important role in this process, resulting in the homogenization of functions and uses in the landscapes of Doñana, thereby affecting the capacity of local ecosystems to provide diversified ecosystem services flows.

This paper adds to recent efforts to transcend traditional natural resource studies previously done in Doñana, in order to delineate landscape management strategies based on the ecosystem services approach (see e.g. Gómez-Baggethun, 2010; Martín-López et al., 2010, 2011; Uhel et al., 2010). Results of the state of the ecosystem services of Doñana show a general picture of moderate deterioration in the case of regulating services. Mean state values seem to be adequate in the provisioning services and rather good in the cultural services. Concerning the trend in the period 1956-2006, the results obtained show levels of deterioration slightly higher than those of the state. Main results obtained through the ecosystem services assessment show a picture of deterioration in the state and trend of the regulating services, certain stability in the provisioning services, and an improvement in the case of the cultural services.

However, important trade-offs can be identified when subcategories of services are analyzed separately based on the scale at which the beneficiaries used them, showing a qualitative shift in the ecosystem services flows. Two aspects of the trade-offs characterizing this shift have been highlighted. First, the enhancement of cultural services that are demanded by stakeholders at national and international scales (e.g., tourism, science, research) at the expense of those that have been historically attached to local stakeholders (e.g., sense of place, local ecological knowledge, gathering, hunting). Second, the improvement in the capacity to provide marketed and high added value services at the expense of non-marketed services. It has been argued that this tendency has promoted the conversion of multi-functional landscapes providing diverse and often non-marketed services into mono-functional landscapes trying to maximize the yield of single marketed services (e.g., strawberry greenhouses, rice crops). Finally, we have pointed out that Doñana might be moving towards a weak sustainability path, as natural capital stocks and functions are being degraded in physical terms while the monetary income obtained from Doñana's natural capital seems to be increasing.

8. Acknowledgements

The authors are grateful to all expert panel members, to S. Sastre, C. Louit and A. Verdú for having provided some cartographical data and to Doñana National and Natural Park for providing facilities to obtain data. This study has been possible with financial support of the Andalusian Ministry of the Environment (project NET413308/1), the Spanish Ministry of Environment (project Doñana 2005-13/2006), and the Autonomic Organism of National Parks (018/2009).

9. References

Balmford, A., Bruner, A., Cooper, P., Costanza, R., Farber, S., Green, R.E., Jenkins, M. & Jefferiss, P. (2002) Economic reasons for conserving wild nature. *Science*, 297, 950-953.

Blondel, J. (2006). The 'Design' of Mediterranean Landscapes: A Millennial Story of Humans and Ecological Systems during the Historic Period. *Human Ecology*, 34, 713–729.

Boyd, J. & Banzhaf, S. (2007) What are ecosystem services? The need for standardized environmental accounting units. *Ecological Economics*, 63, 616-626.

Brandt, J. & Vejre, H. (2002). Multifunctional Landscapes - motives, concepts and perspectives. In: *Multifunctional Landscapes. Volume 1: Theory, Values and History.* Brandt, J. and H. Vejre, (Eds.) 2004 Jessamy, V., Madden, J., Munro, K.., Myers, N., Naeem, S., Paavola, J., Rayment, M., Rosendo, S., Roughgarden, J., Trumper, K. and Turner, R.K., pp. 3-31. WIT press, Southampton, UK.

Butzer, K.W. (2005). Environmental history in the Mediterranean world: cross-disciplinary investigation of cause-and-effect for degradation and soil erosion. *Journal of Archaeological Science*, 32, 1773-1800.

Carpintero, O. (2005). *El metabolismo de la economía española. Recursos naturales y huella ecológica (1955–2000)*. Fundación César Manrique, Islas Canarias (España).

Costanza, R., d'Arge, R., de Groot, R., Farber, S., Grasso, M., Hannon, B., Limburg, K., Naeem, S., O'Neill, R.V., Paruelo, J., Raskin, G.R., Sutton, P. & van der Belt, M. (1997). The value of the world's ecosystem services and natural capital. *Nature*, 387, 253-260.

Custodio, E. (1995). La explotación de las aguas subterráneas y su problemática asociada. In: *VI Simposio de Hidrogeología*: 297-313. Hidrogeología y recursos hidráulicos. v. XX.

Daly, G.C. (Ed.) (1997). *Nature's Services: Societal Dependence on Natural Ecosystems.* Island Press, Washington, DC.

de Groot, R.S. (1992). *Functions of Nature: Evaluation of Nature in Environmental Planning, Management and Decision Making*. Wolters-Noordhoff, Groningen.

de Groot, R.S. (2006). Function analysis and valuation as a tool to assess land use conflicts in planning for sustainable, multi-functional landscapes. *Landscape and Urban Planning*, 75, 175-186.

de Groot, R.S., Wilson, M.A. & Boumans, R.M.J. (2002). A typology for the classification, description and valuation of ecosystem functions, goods, and services. *Ecological Economics*, 41, 393-408.

Douguet, J-M. & O'Connor, M. (2003). Maintaining the integrity of the French terroir: a study of critical natural capital in its cultural context. *Ecological Economics*, 44, 233-254.

Fernández Delgado, C. (2005). Conservation Management of a European Natural Areas: Doñana National Park, Spain. In: *Principles of Conservation Biology*, Groom, M.J., Meffe, G.K. & Carroll, C.R. (Eds.), Sinauer Associates, Inc., Sunderland.

Fisher, B., Turner, R.K. & Morling, P. (2009). Defining and classifying ecosystem services for decision making. *Ecological Economics*, 68, 643–653.

Forman, R. T. T. (1995). *Land Mosaics: The Ecology of Landscapes and Regions*. Cambridge University Press. Cambridge.

García-Novo, F., & Marín-Cabrera, C. (2005). *Doñana, water and biosphere. Doñana 2005*. CHG-MMA, Madrid, Spain.

Gómez-Baggethun, E. (2010). To ecologise economics or to economise ecology: Theoretical issues and operational challenges in ecosystem services valuation. Doctoral dissertation, Universidad Autónoma de Madrid, Spain.

Gómez-Baggethun, E. & Kelemens, E. (2008). Linking institutional change and the flows of ecosystem services. Case studies from Spain and Hungary. In: *Institutional Analysis of Sustainability Problems*, Kluvánková-Oravská, T., Chobotova, V., Jílková, J., (Eds.), pp. 118-145, Slovak Academy of Sciences.

Gómez-Baggethun, E. & de Groot, R. (2010). Natural capital and ecosystem services: the ecological foundation of human society. In: *Ecosystem services, Issues in Environmental Science and Technology 30*, R. E. Hester & R. M. Harrison (Eds.), pp. 118-145, Royal Society of Chemistry, Cambridge.

Gómez-Baggethun, E., Mingorría, S., Reyes-García, V., Calvet, L. & Montes, C. (2010). Traditional ecological knowledge trends in the transition to a market economy: Empirical study in Doñana natural areas. *Conservation Biology*, 24, 721-729.

González Arteaga, J. (1993). *Las marismas del Guadalquivir: etapas de su aprovechamiento económico.* C.P. Antonio Cuevas, Puebla del Río, Sevilla.

González-Arteaga, J. (2005). *El arroz en las marismas del Guadalquivir. Evolución y problemática actual.* Secretariado de publicaciones de la Universidad de Sevilla.

Grove, A.T. & Rackham, O. (2003). The *nature of Mediterranean Europe: an ecological history.* Yale University Press, New Haven.

Hein, L., van Koppen, K., de Groot, R.S. & van Ireland, E.C. (2006). Spatial scales, stakeholders and the valuation of ecosystem services. *Ecological Economics*, 57, 209-228.

Kapp, W., 1983. Social costs in economic development. In: Ullmann, J.E. (Ed.), *Social Costs, Economic development, and Environmental Disruption.* University Press of America, Lanham.

King, R.T. (1966). Wildlife and man. *NY Conservationist*, 20, 9-11.

Klijn, F. & Udo de Haes, H.A. (1994). A hierarchical approach to ecosystems and its implications for ecological land classification. *Landscape Ecology*, 9, 89-104.

Kumar, P. (Ed.) (2010). *The Economics of Ecosystems and Biodiversity, Ecological and Economic Foundations*, Earthscan, London.

Lomas, P.L., Álvarez, S., Rodríguez, M. & Montes., C. (2008). Environmental accounting as a management tool in the Mediterranean context: the Spanish economy during the last 20 years. *Journal of Environmental Management*, 88, 326-347.

MA (Millennium Ecosystem Assessment) (2003). *Ecosystems and their services.* In: *Ecosystems and human well-being*, pp. 49-62, Island Press, Washington DC.

Martín-López, B., Gómez-Baggethun, E., González, J., Lomas, P. & Montes, C. (2009). The assessment of ecosystem services provided by biodiversity: re-thinking concepts and research needs. In: *Handbook of Nature Conservation: Global, Environmental and Economic Issues.*, J. B. Aronoff (Ed.), pp. 261-282, Nova Science Publishers, New York.

Martín-López, B., García-Llorente, M., Gómez-Baggethun, E. & Montes, C. (2010). Evaluación de los servicios de los ecosistemas del sistema socio-ecológico de Doñana. *Forum de Sostenibilidad*, 4, 77-96.

Martín-López, B., García-Llorente, M., Palomo, I. & Montes, C. (2011). The conservation against development paradigm in protected areas: Valuation of ecosystem services in the Doñana social-ecological system (southwestern Spain). *Ecological Economics,* 70, 1481-1491

Menanteau, L. (1984). Evolución Histórica y consecuencias morfológicas de la intervención humana en las zonas húmedas: el caso de las Marismas del Guadalquivir. In: *Zonas Húmedas de Andalucía,* pp. 43-76, Dirección General de Ambiente, MOPU, Madrid.

Montes, C. (2000). The Guadalquivir River Basin and the Doñana Wetlands, Southern Spain - the potential for achieving "Good Water Status" through integrated management of multiple functions and values. In: *Implementing the EU Water Framework Directive: A seminar series on water,* WWF (Ed.), WWF Fresh Water Program, Copenhagen.

Montes, C., Borja, F., Bravo, M. & Moreira, J.M. (1998). *Doñana: una aproximación ecosistémica.* Sevilla, España: Consejería de Medio Ambiente.

Naveh, Z. (2004). Multifunctional, self-organizing biosphere, landscapes and the future of our total human ecosystem. *World futures,* 60, 469-503.

Naveh, Z. & Lieberman, A. (1993). *Landscape ecology: Theory and applications.* Springer-Verlag, New York, USA.

Neumayer, E. (1999). *Weak versus strong sustainability. Exploring the limits of two opposing paradigms.* Cheltenham and Northampton: Edward Elgar Publishing.

Ojeda-Rivera, J.F. (1987). Organización del territorio en Doñana y su entorno próximo. (Almonte). Siglos XVIII-XX. Mº de Agricultura - ICONA. (Monografías, 49). Madrid.

Ojeda-Rivera, J.F. (1990). Doñana cultural landscape. In: *Doñana: La naturaleza en España* (Bilingual ed., VA.AA., pp.18-25, Lunwerg, Barcelona.

Ojeda-Rivera, J.F. (1993). *Doñana: esperando a Godot.* Sevilla, España: Instituto de Desarrollo Regional, Universidad de Sevilla, Sevilla.

Pinto-Correia, T. & Vos, W. (2002). *Multifunctionality in Mediterranean landscapes-past and future.* Proceedings of the Frontis workshop on the future of the European cultural landscapes pp. 135-164. Wageningen, 2002.

Rodríguez Merino, E-E, & Cobo García, D. (2002). Actividades tradicionales. In: *Parque Nacional de Doñana,* García Canseco, V. (Ed.), pp. 354-374 , Canseco Editores, SL. Spain.

Uhel, R, Spyropoulou, R., Breton, F., Beltrame, C., Arévalo, J., Richard, D., Gómez-Baggethun, E., Martín-López, B., Lomas, P., Tomas, P., Ezzine, D., Nichersu, J. & Marin, J. (2010). *Ecosystem accounting and the cost o biodiversity losses: The case of costal Mediterranean wetlands,* European Environmental Agency, TR no 3.

Van der Perk, J. & de Groot, R.S., (2000). *Towards a method to estimate critical natural capital. Discussion paper for the 2nd meeting of the CRITINC-project,* Saint Quentin en Yvelines, Paris, France.

Villa-Díez, A., González-Díaz, J.A. & García-González, J.M. (2000). *Conservación de los valores histórico-artísticos y mantenimiento de las edificaciones públicas: la agricultura de rozas.* Parque Nacional de Doñana, Madrid.

Part 2

Organism-Level Biodiversity

3

Implications of Wood Collecting Activities on Invertebrates Diversity of Conservation Areas

Thokozani Simelane
Centre for African Conservation Ecology, Nelson Mandela Metropolitan University and South African National Parks, Port Elizabeth, South Africa

1. Introduction

Biomass, in the form of deadwood can be described as the end product of series of physiological processes that lead to the deterioration of a piece of wood or the entire tree (Käärik, 1974; Franklin *et al.*, 1987). The rate at which this occurs depend on the exposure of the tree to various physical and physiological stresses (Savory, 1954; Bader *et al*, 1987; Jansson & Jansson, 1995). Once dead, the tree or part thereof can be harvested and used as a source of energy, mostly for cooking or heating in the household. While these are common uses of deadwood, what is also apparent is that deadwood supports ecological systems that are crucial for the maintenance of various components of biodiversity (Graham, 1925; Gosz *et al.*, 1972; Ausmus, 1977; Harmon *et al.*, 1993). As a result natural processes and systems of deadwood production which are often well preserved and maintained within the conserved environment (Graham, 1925; Raphael & Morrison, 1987; Harmon *et al.*, 1993) requires that deadwood be regarded as critical part of biodiversity management (Harmon *et al.*, 1993; Bergeron, 2000; Andrzej, 2002; Hagar, 2007).

In the past years, the management of deadwood within conservation areas has solely been based on observations that 1) deadwood provides habitat for different species of birds, bats and mammals (Brandlmaier *et al.*, 2004), 2) deadwood serves as a source of food for various organisms (Raphael & White, 1984; Harmon *et al.*, 1993) including the less visible invertebrates, fungi and lichens and that 3) deadwood has a potential of supplementing soil organic nutrient (Hart, 1999) and thus promoting soil fertility.

As with the case of standing dead trees (Andrzej, 2002), which are used by different vertebrates, such as birds for nesting sites (Johnston & Odum, 1956; Du Plessis, 1995), fallen dead trees are used by small mammals (Rhoades, 1986), reptiles and various species of invertebrates as mating sites, shelter or sources of food (Hirth, 1959; Harmon 1982). All these observations, combined have increased the value of deadwood as playing a key role in sustaining the efficiency and productivity of the ecological systems within conservation areas.

Unfortunately in most parts of the world deadwood still remain the main source of energy and is in great demand for domestic fuel. This is the main cause for concern among conservation agencies (Anderson & Fishwick, 1984; Wall & Reid, 1993; Abbot & Mace, 1999) as it poses a threat to biodiversity that is housed within deadwood (Kavin, 2001). Of considerable importance is that among certain societies dead wood is not only used for energy alone but has some cultural link (Furness, 1979). An example is the Xhosa, Vhenda and Zulu communities of South Africa where deadwood is specifically collected and used during traditional functions such as weddings, funerals and circumcision ceremonies (Furness, 1979). The combined effect of this has resulted in the decline of the availability of deadwood outside conservation areas (Shackleton, 1993a; Wall & Reid, 1993; Maruzane & Cutler, 2002). This has placed pressure on conservation areas to make this resource available to communities (Anderson & Fishwick, 1984; Bembridge, 1990). With the possible negative effects associated with deadwood harvesting, it is clear that the collection of deadwood from conserved areas might disturb and fragment some ecosystem processes and this could increase species loss and extinction.

The debate on deadwood availability outside conservation areas has largely been limited to its shortage as caused by over-harvesting and demand (Arnold, 1978; Anderson & Fishwick, 1984; Bembridge & Tarlton, 1990; Shackleton, 1993b) with its exploitation being reported as leading to habitat destruction for wood-inhabiting organisms and deforestation (Mainguet, 1991). Little attention has been given to the ecological effects of deadwood harvesting or the role of deadwood in maintaining ecological integrity and biodiversity (Banerjee, 1967; Bilby, 1981; Bilby & Likens, 1980) outside conservation areas.

This oversight is despite the well-recognized fact that the presence of wood-inhabiting organisms in deadwood attracts other organisms that are either predators of these organisms or their larvae (Fager, 1968; Harmon *et al.*, 1993). This relationship has long been recognized and appreciated by entomologists and has generated some interest in research and management of biodiversity associated with deadwood (Graham, 1925; Fager, 1968; Käärik, 1974; Deyrup, 1981; Araya, 1993; Bennett *et al.*, 1994; Lachat *et al.*, 2006). Such plant-animal interactions has been identified as one of the dominant biotic interactions (Graham, 1925; Farrell *et al.*, 1992) that sustains much of the terrestrial faunal diversity (Samways, 1993) through the support of ecological interactions that exist among terrestrial living organisms. Thus, activities such as collection of deadwood for energy from conservation areas may indirectly affect the maintenance of these interactions, and hence the conservation of the diversity of organisms that are associated with deadwood (Gandar, 1984; Anderson & Fishwick, 1984).

To highlight some of these threats and their possible effects on biodiversity, invertebrate diversity associated with deadwood was determined through an experimental study that was conducted in Vaalbos National Park (VNP, South Africa). The investigation addressed the hypothesis that the collection of deadwood for energy from conservation areas does not only endanger trees but also other elements of biodiversity. These may include those invertebrates whose existence is largely dependent on the presence of deadwood. In investigating this, it was hypothesized that the invertebrate diversity associated with deadwood correlate with the increasing wood size, and hence the value of the material as both fuelwood and in supporting biodiversity.

2. Materials and methods

Invertebrates in deadwood were harvested using the following procedure. Deadwood from a range of unidentified plant species of the park was randomly collected from three selected sites in the park, simulating the method of harvesting deadwood by communities and transported to a research station where the invertebrates were extracted from the deadwood.

As wood collectors prefer wood size that can be easily transported by hand (Bembridge & Tarlton, 1990), three deadwood sizes (i.e. Finger size (FS) (<2 cm in diameter), Arm-size (AS) (2 - 5 cm diameter), and Leg-size (LS) (> 5 cm diameter but less than 10 cm) were identified and chosen for the study. These were also regarded as representing wood pieces that break and burn easily (Bembridge & Tarlton, 1990).

Collected deadwood was cut into 30 cm long pieces, weighed and loaded into 18 modified 100-litre drums that were divided into replicates of each wood size. The drums were modified such that the bottom one third of the drum was separated from the upper portion by a 38-mm mesh grid supported by iron bars. The lower separated portion was used as pitfall trap in which emerging invertebrates were collected. Each pitfall trap was filled with 5 litres of water that prevented invertebrates from leaving the trap.

Twelve of the drums served as an "illuminated" insect harvest, with each wood size class having four replicates. Drums were illuminated by 60 watt white light bulbs that were suspended 60 cm above the wood layer. This encouraged the mobility of the invertebrates to make them leave the wood. The lights were connected to a photo-sensor switch, which followed a reserve diurnal cycle to ensure 24-hour lighting so as to maintain light throughout the period of the experiment.

The six remaining drums (two replicates for each wood size class) were left without light and represented the uncontrolled condition without apparently induced invertebrate mobility. The top of each drum was covered with black cloth that ensured that sunlight did not interfere with the harvest process and that insects did not escape from or enter the drums from the outside. All drums were placed in the shade to reduce temperature variations during the experiment. The invertebrate harvest was conducted over two time periods, both during the summer months and both running for a period of nine months. These periods were selected because the activity of invertebrates is recognizably high during this period of the year (Davies, 1994).

Invertebrates were collected from the bases of the drums once a week, preserved in 70% alcohol and identified to family level (Davies, 1994). The families were categorized into the following functional guilds: obligate wood dwellers (OWD), semi-obligate wood dwellers (SOWD) and associates of deadwood (AODW), depending on their level of association with deadwood. After the experiment was completed, the deadwood was broken down with a chisel and hammer to determine whether any invertebrates remained within the wood. Invertebrates collected through this method were added to the sample of emerged invertebrates.

2.1 Statistical analysis

Data collected from the two seasons in which the experiment was conducted and from illuminated and non-illuminated drums were first tested for statistical differences. Where there was no statistical differences, data were pooled and analyzed together. Where there was a statistical difference, data were analyzed separately (e.g. numbers of invertebrates collected from illuminated drums with LS wood). The differences between numbers of invertebrates collected from illuminated and non-illuminated drums were compared statistically using one-way Kruskal-Wallis Analysis of Variance (Zar, 1984). Differences in a numbers of invertebrates and the larvae collected from three wood classes were also compared statistically using a one-way Kruskal-Wallis Analysis of Variance (Zar, 1984).

3. Results

In analysing and interpreting the results, it was considered that environmental factors, such as humidity, temperature and weather might have played a role in influencing the emergence of invertebrates from the wood. However, the fact that the drums were housed in the same conditions negated this concern. While attempts were made to identify all collected invertebrates into families some such as Pseudoscorpionida and Lepidoptera were identified to Order level only, this was due to a limited ability available to identify these invertebrates further.

The sequence of emergence of invertebrates from deadwood was such that the buprestids and cyrambecids were the first to emerge, while groups such as clerids and halictids (Table 1) emerged at the later stages of the experiments. One thousand seven hundred and fifty invertebrates were collected (Table 2). For two of the wood size classes (FS ($H = 3.71$, $df = 5$, $p>0.05$) and AS ($H = 4.56$, $df = 5$, $p > 0.05$) there was no statistically significant difference between the invertebrates collected from illuminated and non-illuminated drums (Table 2). For the leg size wood class, the illuminated drums yielded a significantly higher ($H = 23.24$, $df = 5$, $p < 0.001$) number of invertebrates than drums without light (Table 2).

An average of 1.5 ± 2.3 (Average ± SD) invertebrates per kilogram of FS wood, 2.5 ± 3.1 per kilogram of AS wood and 4.5 ± 5.6 per kilogram of LS wood (Figure 1) were harvested from each size class of wood. This was interpreted as indicating that a head-load of deadwood (Bembridge & Tarlton, 1990, Shackleton, 1993b) with an approximate mass of 20 kg of finger-size wood could contain an average of 30 ± 1.4 invertebrates, a head-load of arm-size wood could contain an average of 50 ± 2.7 invertebrates and a head-load of leg-size wood 90 ± 1.5 invertebrates of a variety of guilds, families and species.

3.1 Invertebrate guilds associated with deadwoods

The collected invertebrate fell into three broad functional guilds i.e. obligate wood dwellers (OWD), semi obligate wood dwellers (SOWD) and associate of dead woods (AODW) (Table 2). This classification was based on taxonomic categorization; feeding behavior and the maximum time an invertebrate was found to spend in deadwood (Scholtz & Holm, 1996). Nine of the identified families i.e. 26 % of the total numbers of families collected were identified as obligate wood dwellers (OWD) (Appendix). These invertebrates spend their entire lifecycle in deadwood. They inhabit the tree while it is still alive, with certain stages of their development (larval stage) being completed in dead wood (Harman *et al.*, 1993). This group has a considerable pathological effect on trees and influence tree mortality (Harmon *et al.*, 1993). The Halictidae (46.5 % of the total number of OWD collected), Buprestidae (25.1 %) and Cerambycidae (22.9 %) dominated this group.

The Pseudoscorpionidae (Order) and 14 (40 % of the total number of families collected) other identified families (Appendix) were classified as a group that depends on deadwood for only certain of their lives (Table 1). This group was referred to as semi-obligate wood dwellers (SOWD) and spends a portion of their lives in deadwood. They are either predators of OWD invertebrates (e.g. Histeridae), colonize holes excavated by the larvae of OWD group (e.g Carabidae) or are parasitoids (e.g. Chalcididae and Gasteruptidae) of these larvae. The dominant families that represented this group were Formicidae (20.4 %) of the total number of SOWD collected), Histeridae (15.6 %) and Lepismatidae (14.7 %).

Lepidoptera (Order) and 13 other families (33 % of the total number of families collected) were identified as those invertebrates that use deadwood either for hunting, hiding or feeding on fungi that grow on the deadwood. This group was referred to as associates of deadwood (AODW) (Scholtz & Holm, 1996) (Appendix). These invertebrates can survive and complete their life cycle in the absence of deadwood (Scholtz & Holm, 1996)

(Appendix). Megachilidae (21.2 %), Galleridae (11.9 %) and Tenebrionidae (11.0 %) represented this category.

Order	Family	FS			AS			LS		
		Non-illuminated	Illuminated	Total Mean	Non-illuminated	Illuminated	Total Mean	Non-illuminated	Illuminated	Total Mean
Coleoptera	Cerambycidae	2.3±1.3	6.3±5.1	8.6	19.0±7.1	25.0±12.2	44.0	7.5±3.5	22.5±17.1	30.0
Coleoptera	Buprestidae	0.5±0.4	0.5±1.0	1.0	21.0±11.3	20.3±10.1	41.3	9.5±0.7	37.0±14.2	46.5
Coleoptera	Bostrychidae	0.00	0.5±1.0	0.5	0.00	0.5±1.0	0.5	0.00	3.0±3.5	3.0
Coleoptera	Lyctidae	0.00	0.00	0.0	0.00	0.00	0.0	0.00	0.5±1.0	0.5
Coleoptera	Mordelidae	0.00	0.00	0.0	0.00	0.00	0.0	1.3±2.4	0.8±1.5	2.1
Coleoptera	Anobiidae	0.00	0.00	0.0	0.00	1.0±2.0	1.0	0.00	2.8±2.5	2.8
Coleoptera	Cleridae	0.00	0.00	0.0	0.00	0.00	0.0	0.3±1.3	2.8±2.5	3.1
Coleoptera	Orussidae	0.00	0.00	0.0	0.00	0.00	0.0	0.00	0.3±0.5	0.3
Hymenoptera	Halictidae	2.5±3.5	1.3±0.5	3.8	17.0±7.1	3.2±1.4	20.2	0.00	82.0±41.4	82.0
Coleoptera	Histeridae	0.00	0.00	0.0	0.00	0.00	0.0	0.00	1.0±2.0	1.0
Coleoptera	Carabidae	6.5±3.5	2.7±1.4	9.2	0.00	4.0±6.1	4.0	0.00	4.3±1.8	4.3
Hemiptera	Aradidae	0.00	0.00	0.0	0.00	0.8±0.9	0.6	0.00	0.5±1.0	0.5
Coleoptera	Elateridae	0.00	0.00	0.0	0.00	0.00	0.0	0.5±1.3	3.0±3.5	3.5
Hymenoptera	Colletidae	0.00	0.00	0.0	0.00	0.00	0.0	0.00	1.5±3.0	1.5
Hymenoptera	Chrysididae	0.00	0.00	0.0	0.00	0.5±0.9	0.5	0.00	2.0±2.4	2.0
Hymenoptera	Chalididae	0.00	0.00	0.0	0.5±1.4	0.8±0.3	1.3	0.00	1.3±2.5	1.3
Coleoptera	Curculionidae	0.00	0.00	0.0	0.00	0.00	0.0	0.00	1.3±1.5	1.3
Hymenoptera	Gasteruptidae	0.00	0.00	0.0	0.5±2.3	1.0±1.4	1.5	0.00	3.5±2.3	3.5
Hymenoptera	Sphecidae	0.00	0.00	0.0	1.2±0.4	3.3±4.7	3.5	0.00	3.3±1.9	3.3
Hymenoptera	Chalcidoidae	0.00	0.00	0.0	3.7±0.3	4.3±6.6	8.0	0.00	0.3±0.5	0.3
Hymenoptera	Formicidae	0.00	0.00	0.0	4.2±1.6	8.8±4.7	13.0	3.0±4.2	8.3±3.5	11.3
Thysanura	Lepismatidae	0.00	0.00	0.0	1.4±2.6	7.5±4.2	8.9	3.2±2.3	9.5±6.1	12.7
Pseudoscorpionida		0.00	0.8±1.5	0.8	1.5±2.1	5.0±0.2	6.5	0.00	11.3±8.8	11.3
Hymenoptera	Megachilidae	0.00	0.00	0.0	2.5±3.5	18.5±1.0	21.0	2.0±2.8	2.5±2.1	4.5
Hemiptera	Coreidae	0.00	0.00	0.0	0.00	0.00	0.0	0.00	1.0±2.0	1.0
Hemiptera	Pyrrhociridae	0.00	0.00	0.0	0.00	0.00	0.0	0.00	1.5±1.2	1.5
Lepidoptera	Galleriidae	0.00	0.00	0.0	0.00	0.00	0.0	0.00	1.5±1.7	1.5
Coleoptera	Chrysomelidae	0.00	0.00	0.0	0.00	2.3±0.5	2.3	0.00	1.5±2.3	1.5
Hemiptera	Cicadellidae	0.00	0.00	0.0	0.00	0.00	0.0	0.00	1.5±3.0	1.5
Coleoptera	Tenebrionidae	0.00	0.00	0.0	0.00	1.5±1.5	1.5	0.5±1.3	2.5±2.3	3.0
Hemiptera	Pentatomidae	0.00	0.00	0.0	0.5±1.3	1.2±0.3	1.7	0.00	0.8±1.5	0.8
Phasmotodea	Phasmotidae	0.00	0.00	0.0	0.5±1.5	0.00	0.5	0.00	1.0±1.2	1.0
Blattodea	Blattidae	0.00	0.00	0.0	0.6±1.3	0.5±0.5	1.1	1.0±1.4	2.3±1.7	3.3
Lepidoptera		0.00	0.00	0.0	0.00	0.00	0.0	0.00	2.0±1.7	2.0
Mantodea	Mantidea	1.2±0.2	2.8±3.2	4.0	0.00	0.00	0.0	0.00	0.0	0.0
Orthoptera	Gryllidae	0.00	0.00	0.0	0.5±1.2	0.5±1.2	1.0	0.00	0.5±1.0	0.5
		5	7		15	21		10	35	

Table 1. Mean number / kg (± SD) of invertebrates collected from three different sizes of deadwood (FS = finger size; AS = arm size, LS = leg size; OWD = Obligate wood dwellers; SOWD = Semi-obligate wood dwellers; AODW = Associate of dead wood). Invertebrates are arranged according to the sequence of emergence from the wood.

Taxon	Guild	FS			AS			LS			
		Illuminated	Non Illuminated	Total	Illuminated	Non-illuminated	Total	Illuminated	Non-illuminated	Total	Total
Cerambycidae	OWD	16	9	25	97	41	138	80	25	105	268
Buprestidae	OWD	2	1	3	67	56	123	154	13	167	293
Bostrychidae	OWD	2	0	2	2	0	2	12	0	12	16
Lyctidae	OWD	0	0	0	0	0	0	2	0	2	2
Mordelidae	OWD	0	0	0	0	0	0	3	0	3	3
Anobidae	OWD	0	0	0	4	0	4	8	3	11	15
Cleridae	OWD	0	0	0	7	3	10	14	2	16	26
Orissidae	OWD	0	0	0	0	0	0	1	0	1	1
Halictidae	OWD	32	8	40	123	21	144	358	0	358	542
Sub Total		52	18	70	300	121	421	632	43	675	1166
Histeridae	SOWD	0	0	0	0	0	0	4	0	4	4
Carabidae	SOWD	28	11	39	16	0	16	17	0	17	72
Aradidae	SOWD	0	0	0	3	0	3	2	0	0	5
Elateridae	SOWD	0	0	0	0	0	0	10	2	12	12
Colletidae	SOWD	0	0	0	0	0	0	5	0	5	5
Chrysididae	SOWD	0	0	0	4	0	4	8	0	8	12
Chalicididae	SOWD	0	0	0	2	1	3	5	0	5	8
Curculionidae	SOWD	0	0	0	0	0	0	5	0	5	5
Gasteruptidae	SOWD	0	0	0	3	1	4	17	0	17	21
Sphecidae	SOWD	0	0	0	10	3	13	16	0	16	29
Chalcidoidae	SOWD	0	0	0	12	17	29	1	0	1	30
Formicidae	SOWD	0	0	0	43	12	55	33	6	39	94
Lepismatidae	SOWD	0	0	0	22	8	30	31	7	38	68
Pseudoscorpionidae	SOWD	3	0	3	37	6	43	55	0	55	101
Sub Total		21	11	42	152	48	200	209	15	224	466
Megachilidae	AODW	0	0	0	11	0	11	10	4	14	25
Coreidae	AODW	0	0	0	0	0	0	4	0	4	4
Pyrrhociridae	AODW	0	0	0	0	0	0	6	0	6	6
Chrysomelidae	AODW	0	0	0	0	0	0	6	0	6	6
Cicadellidae	AODW	0	0	0	0	0	0	6	0	6	6
Tenebrionidae	AODW	0	0	0	2	1	3	8	2	10	13
Colletidae	AODW	1	1	1	0	0	0	0	0	0	2
Pentatomidae	AODW	0	0	0	0	0	0	3	0	3	3
Phasmotodea	AODW	0	0	0	2	2	4	4	0	4	8
Blattidae	AODW	0	0	0	3	3	6	9	2	11	17
Lepidoptera	AODW	0	0	0	0	0	0	6	0	6	6
Mantodea	AODW	6	5	11	0	0	0	0	0	0	11
Gryllidae	AODW	0	0	0	1	1	2	2	0	2	4
Sub Total		7	6	13	27	7	34	70	8	78	125
Total		90	35	125	479	176	655	911	66	977	1757

Table 2. Numbers (per kg) of invertebrates collected from the studied wood sizes. (FS = finger size; AS = arm size; LS = leg size; OWD = Obligate wood dwellers, SOWD = Semi-obligate wood dwellers, AODW = Associate of deadwood).

3.2 Deadwood diameter and invertebrate assemblage

Wood with larger diameter (AS and LS classes) were found to have a significantly higher number ($H = 34.3$, $df = 2$, $p < 0.001$) of invertebrates than those with a smaller diameter (finger size (<2cm) (Figure 1, Table 2). This was understood to be due to the size of the niche provided by this wood class.

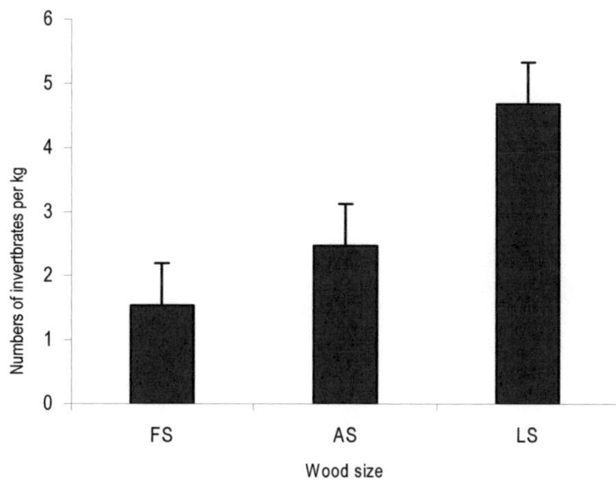

Fig. 1. Average numbers (±SD) of invertebrates recorded as occurring in a kilogram of the tree studied deadwood sizes. (FS = finger sizes, AS = arm size, LS = Leg size of deadwood).

Both arm- (AS) and leg-size (LS) wood classes had the higher numbers of the size-specific invertebrates (invertebrates limited to wood of specific diameter) (Table 2), with some invertebrates only occurring in wood of the largest diameter (LS) (Table 2). The diversity of invertebrates (i.e. number of families per wood size) calculated as occurring in a kilogram of each wood size did not differ significantly ($H = 0.00$, $df = 36$, $p > 0.05$) between the three studied wood sizes (Figure 3).

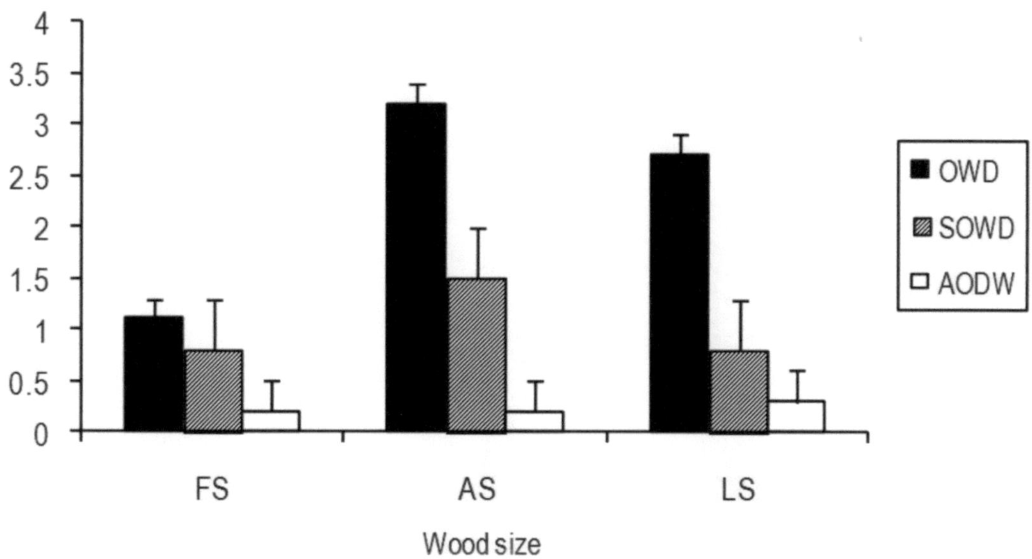

Fig. 2. Average number (±SD) of invertebrates from each category of invertebrates recorded as occurring in a kilogram of three studied wood sizes. (FS = Finger size, AS = Arm size, LS = Leg size).

In addition to adult invertebrates, an average of 13.2 ± 1.5 larvae per kg and 7.4 ± 0.6 larvae per kg (through breaking wood) were collected. Collected larvae were identified as belonging to four taxa (Table 3). Three of the families were those of OWD (Buprestidae, Cerambycidae and Cleridae) and one for the AODW (Lepidoptera (Order)(Table 3). Larvae for buprestids (74.5% of total collected larvae) and Cerambycids (12.8 % of total collected larvae) were significantly ($H = 6.12$, $df = 4$, $p < 0.01$) more abundant than those of Cleridae (10.6%) and Lepidoptera (2.1%).

Taxon	FS	AS	LS
	Average mass (g)	Average mass(g)	Average mass(g)
Buprestidae	0.001±3.4 (n = 9)	0.11±2.45(n = 33)	0.14±4.53(n = 63)
Cerambycidae	0.02±1.98(n = 5)	0.12±1.35(n = 7)	0.23±3.54(n = 6)
Cleridae	0.01±4.67(n = 2)	0.01±1.34(n = 5)	0.10±2.11(n = 8)
Lepidoptera	0.0 (n = 0)	0.13(n = 1)	0.13±3.59(n = 2)

Table 3. Total numbers and average (±SD) mass of larvae collected from three different sizes of deadwood (FS = finger size, AS = arm size and LS = leg size).

Larvae were more abundant in larger diameter wood than in smaller diameter wood ($H = 3.8$, $df = 2$, $p < 0.05$). Notably, larvae that occurred in all three sizes of deadwood differed in body size ($H = 5.7$, $df = 3$, $p < 0.01$) (Table 3), with larvae from deadwood of larger diameter (AS and LS) having higher average mass than those from deadwood with smaller diameter (finger-size) (Table 3).

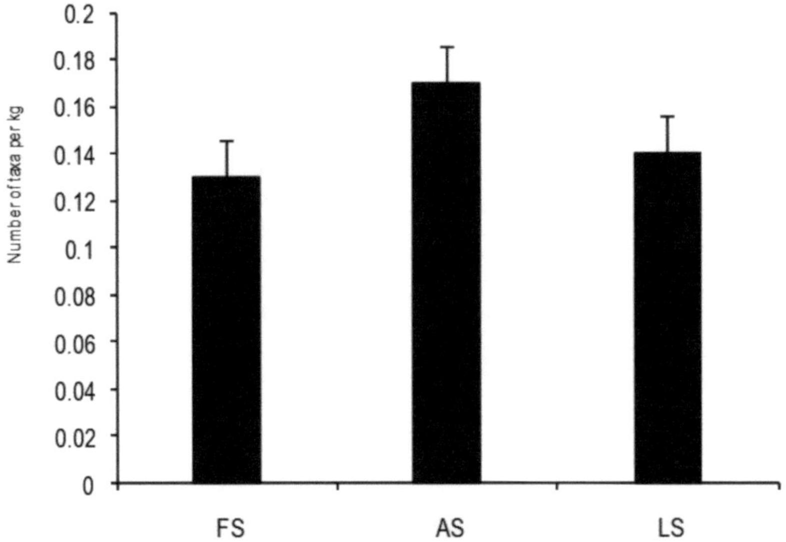

Fig. 3. Average numbers (±SD) of taxa recorded as occurring in a kilogram of three sampled wood sizes. (FS = finger size, AS = arm size, LS = leg size).

4. Discussion

This study shows that deadwood supports a broad diversity of invertebrates. These belong to a variety of guilds (Deyrup, 1976) and types (Graham, 1925) and differ in numbers (Deyrup, 1981; Harcombe & Marks, 1983) within the different sizes of deadwood (Fager, 1968; Harmon, 1982; Marshall, Setälä & Trofymow, 1998). Of notable significance is that, while Deyrup (1981) recorded more than 300 species of invertebrates from single species of Douglas-fir, this study recorded 1 757 individuals of invertebrates, identified as belonging to thirty-six families (Table 2). With such a high number of invertebrates species recorded and the wide variety of taxa found to be associated with deadwood, it is obvious that different tree species, although in different stages of their developments serve as a host to a diversity of invertebrate species (Saniga & Schütz, 2001). The fact that each stage of the tree is associated with a particular community of invertebrates (Araya, 1993; Bennett *et al.*, 1994), indicates that a thorough investigation of the role and contribution of deadwood to the conservation of biodiversity needs to be investigated further to determine the other cryptic implications of collecting deadwood on biodiversity of conservation areas.

What this chapter highlights which is of critical importance in respect to wood inhabiting invertebrates and the conservation of invertebrate diversity through maintenance of deadwood in conservation areas, is that some invertebrates are distinctly characterized of and limited to the habitat that is only provided by deadwood (Brues, 1920; Deyrup, 1976). This is obvious for the OWD and SOWD guilds (Käärik, 1974; Ausmus, 1977) whose life history is confined within deadwood such that these invertebrates cannot survive in the absence of deadwood (Brues, 1920; Brumwell, Craig & Scudder, 1998) (Table 1). This indicates that the removal of deadwood from conservation areas could have direct negative effects on those organisms that rely on the presence of deadwood for survival (Blanchet & Shaw, 1978; Baker, 1979).

As each part (Deyrup, 1981) and size of wood is distinctly associated with different groups of invertebrates (Baumbeger, 1919; Deyrup, 1981) that colonize trees at different levels of decay (Christensen, 1984; Gashwiler, 1970), it is obvious that the removal of trees from conserved systems may interrupt the processes of ecological succession that takes place in dying or dead trees (Saunders, Hobbs & Margules, 1991; Harmon *et al.*, 1993; Sánchez-Azofeifa *et al.*, 1999). As these processes are associated with chemical changes that take place in a senescing tree, this would thus impede the progression of invertebrate from one group (e.g. truly wood eating (xylophagous) invertebrates (OWD) to those that are able to digest wood into fine powder (e.g. Lyctidae) (Deyrup, 1981). This progression is critical for the maintenance of the natural production of deadwood in a protected ecosystem. For example, true wood-eating invertebrates (xylophagous), with their ability to digest and assimilate food material from fresh wood tissues (Graham, 1925; Hickins, 1963; Käärik, 1974), trigger the death of the tree. Without this group, potential food material in wood can be locked up and the development of the succeeding stages of wood decay would be impeded such that the entire process of deadwood production would be retarded. This would normally lead to a scarcity of deadwood and would, in turn, trigger the destructive harvesting of wood through the cutting of live trees (Anderson & Fishwick, 1984; Gandar, 1984).

This process would then normally lead to vegetation clearing which is prevalent in unprotected areas. The evidence provided by this study suggest that it will be necessary to give serious consideration to all the effects associated with the removal of deadwood from conservation areas. Such effects may have long-term negative implications that would directly affect the biodiversity associated with deadwood.

This study has identified a group of wood-dwelling invertebrates that would be potentially vulnerable to habitat loss and population decline in the event of wood collection from conservation areas if deadwood harvesting is considered. It is therefore recommended that studies be undertaken to measure the impact of various proportions of wood being removed, and that the consequences of wood removal on this element of biodiversity and the processes provided by these species be monitored.

As the replacement of deadwood takes a long time, it is also obvious that the impacts associated with the removal of deadwood from conservation areas would have a long term affects and may have extended effects on those organisms that depend on the presence of deadwood for survival (Graham, 1925; Holmes & Sturges, 1975; du Plessis, 1995). These include woodpeckers, snakes and different species of reptile that colonize deadwood killed by wood inhabiting invertebrates (Elton, 1966; Fager, 1968; Losey & Vaughan, 2006).

In addition, as the presence of wood-inhabiting invertebrates attracts other organisms to wood, either as predators, parasitoids or through symbiotic relationships (Graham, 1925; Johnston & Odum, 1956; Conner, Miller & Adkisson, 1976; Mannan, Meslow & Weight, 1980; Bader, Jansson & Jansson, 1995), the removal of wood from conservation areas would limit this diversity of organisms (Hirth, 1959; Hamilton, 1978; Manna, Meslow & Weight, 1980; Farrell, Milter & Futuyma, 1992). Thus, maintaining the presence of deadwood as part of the ecosystem of conservation areas seem to enhance the success of conservation areas in conserving biodiversity (Brumwell, Craig & Scudder, 1998).

In conclusion, it could be mentioned that in the absence of firm evidence of the amount of wood that can be collected from conservation areas without incurring negative effects on the web of biodiversity associated with deadwood, it is difficult to commend wood harvesting from conservation areas as being sustainable. This calls for increased efforts towards developing an understanding of the importance of deadwood in mantaining biodiversity within protected ecosystems. This should include the development of methods of harvesting deadwood from conservation areas with little effects on biodiversity.

What is emerging is that deadwood (especially in Europe) is gaining much recognition as the indicator of ecosystem health such that in various parts of Europe researchers and government authorities have started to survey the role of deadwood in natural forests (Sippola *et al.*, 1998; Brandlmaier *et al.*, 2004). The aim of these studies is to determine how much deadwood should be mantained in natural forest so as to manage healthy forest ecosystem. Initiatives like these need to be extended to other areas sush as Africa where the use and demand for deadwood far exceeds production.

5. Appendix

Families of invertebrates collected from deadwood and the reasons for their association with deadwood. Reasons were extracted from Scholtz & Holm (1996).

Taxon	Guild	Reason
Cerambycidae	OWD	Larvae are wood borers.
Buprestidae	OWD	Adults attack moribund (i.e. dying) rather than dead wood, larvae are woodborers.
Bostrychidae	OWD	Both adult and larvae are woodborers.
Lyctidae	OWD	Both adult and larvae are wood borers, with larvae reducing wood to fine powder.

Taxon	Guild	Reason
Mordelidae	OWD	Larvae feed in live and decaying wood.
Anobiidae	OWD	Larvae bore in the wood and bark of dead trees.
Cleridae	OWD	Predaceous upon other insects, predominant food being larvae of lignicolous beetles.
Orussidae	OWD	Larval parasitoids of wood boring beetles of the buprestids and Cerambycids
Halictidae	SOWD	They nest in burrows either in the ground or less commonly in wood.
Histeridae	SOWD	Both the adults and larvae prey on the larvae of other insects.
Carabidae	SOWD	Predaceous with some noted to live in decaying plant material such as logs and leaf litter
Aradidae	SOWD	Mycetophagous, found under loose bark of dead branches feeding on fungi.
Elteridae	SOWD	Adults feed on vegetable matter such as leaves, flower petals or pollen.
Chrysididae	SOWD	Larvae are external parasites of the fully fed or immature insects.
Chalcididae	SOWD	Secondary parasitoids, which attck larvae or pupae of large variety of insects.
Chalcidoidae	SOWD	Some parasitic, others phytophagous and others hyperprasitoids.
Curculionidae	SOWD	Most are phytophagous
Gasteruptidae	SOWD	Parasitic in the nest of solitary wasps and bees, especially those that nest on dead wood.
Pseudoscorpionida	SOWD	Widely distributed in various habitats, commonly under the bark of deadwood.
Sphecidae	SOWD	Most are predators and prey on a variety of insects
Lepistmatidae	SOWD	Occupy a variety of habitats including houses.
Megachilidae	AODW	Pollen collecting. Nest in burrows excavated by larvae of wood boring beetles.
Colletidae	AODW	Nest on pithy plant stems or in existing burrows in wood excavated by larvae of wood boring beetles.
Coreidae	AODW	Phytophagous, attack young plant shoots.
Gryllidae	AODW	Most species are omnivorous and nocturnal.
Colletidae	AODW	Their nests are usually made wither burrowing into the ground or utilizing existing burrows in wood such as those made by wood boring beetle larvae.
Pyrrhocoridae	AODW	Phytophagous. They are the main transmitters of nematospora fungi.
Galleriidae	AODW	Larvae feed on a variety of dried substances.
Chrysomelidae	AODW	Adults feed on plants but are also adapted to different types of life.
Cicadellidae	AODW	Most types of vegetation serve as a host, often abundant on shrubs and trees.
Tenebrionidae	AODW	Some are phytophagous with larvae living in decaying wood and plant litter.
Blattidae	AODW	Often found around areas where humans live.
Lepidoptera	AODW	Adult feed entirely on nectar, over ripe fruit and other liquid substances.
Montodea	AODW	Often solitary, occurring mostly on vegetation and use deadwood as hunting grounds.
Pentatomidea	AODW	Include a number of pests that are of economic importance. Use deadwood for refuges.
Phasmatidae	AODW	May be common in dry grass, which they resemble. Use deadwood as refuges.

6. References

Abbot, J. I. O. & Mace, R. (1999). Managing protected woodlands fuelwood collection and law enforcement in Lake Malawi National Park. *Conservation Biology* 13: 518 - 421.

Anderson, D. & Fishwick, R. (1984). *Fuelwood consumption and deforestation in African countries*. World Bank staff's Working paper No 704. Washington DC.

Araya, K. (1993). Relationship between decay types of deadwood and occurrence of lucanid beetles (Cleoptera: Lucanidae). *Applied Entomological Zoology* 28: 27-33.

Arnold, J. E. M. (1978). *Wood energy and rural communities*. Paper presented at the 8th world Forestry Congress. Jakarta. Indonesia.

Ausmus, B. S. (1977). Regulation of wood decomposition rates by arthropod and annelid populations. *Ecological bulletin* 25: 180-192

Bader, P., Jansson, S. & Jansson, B. G. (1995). Wood inhabiting fungi and substratum decline in selectively logged boreal spruce forests. *Biological Conservation* 72: 355 - 362.

Baker, C. O. (1978). *The impacts of log jam removal on fish populations and stream habitat in western Oregon*. Msc thesis, Oregon State Univ. Colorado.

Baumbeger, J. P. (1919). A nutritional study of insects with special reference to micro-organisms and their substrata. *Journal of experimental Zoology* 28: 1-81.

Barnerjee, B. (1967). Seasonal changes in the distribution of millipede Cylindroiulus panctatus (Leach) in decaying logs and soil. *Journal of Animal Ecology* 36: 171-177.

Bembridge, T.J. (1990). Woodlots, woodfuel and energy strategies for Ciskei. *South African journal of forestry* 155: 42-50.

Bembridge, T. J. & Tarlton J. E. (1990). Woodfuel in Ciskei: A headload study. *South African Journal of Science* 54: 88-93.

Bennett, A. F., Lumsden, L. F. & Nicholls A. O. (1994). Tree hollows as a resource for wildlife in remnant woodlands: spatial and temporal patterns across the northern plains of Victoria, Australia. *Pacific Conservation Biology* 1: 222-235.

Bergerron, Y. (2000). Species and stand dynamics in mixed woods of Quebec's southern boreal forest. *Ecology* 81: 1500-1516.

Bilby, R. E. & Lickens, G. E. (1980). Importance of organic debris dams in the structure and function of stream ecosystem. *Ecology* 61: 1234-1243.

Bilby, R. E. (1981). Role of organic debris dams in regulating the export of dissolved and particulate matter from a forested watershed. *Ecology* 61: 1234-1243.

Blanchett, R. A. & Shaw, C. G. (1978). Associations among bacteria, yeasts and basidiomycetes during wood decay. *Phytopaththology* 68: 631-637.

Brandlmaier, H., Steindlegger, G., & Pollard, D (eds). (2004). *Deadwood-living forests*. WWF Report. 19pp.

Brumwell, L. J., Craig, K. G. & Scudder, G. G. (1998). Litter spiders and carabid in successional Douglas-fir in British Columbia. *Northwest Science* 72(2): 94pp.

Brues, C. T. (1920). The selection of food plants by insects. *The American Naturalist* 54: 313-332.

Conner, R. N., Miller, O. K. & Adkinsson, S. (1976). Woodpecker dependence on trees infected by fungal heart rots. *The Wilson Bulletin* 88(4): 575-581.

Christensen, O. (1984). The states of decay of woody litter determined by relative density. *Oikos* 42:211-219.

Davies, A. L. V. (1994). Community organization in a South African, winter rainfall, dung beetle assemblage (Cleoptera:Scarabaeidae). *Acta Oecologia* 15: 727-738.

Deyrup, M. A. (1976). *The insect community of dead and dying Douglas-fir: Diptera, Coleptera and Neuroptera*. PhD thesis, Univ. of Washington, Seattle.

Deyrup, M. A. (1981). Deadwood decomposers. *Natural History* 90:84-91.

Du Plessis, A. M. (1995). The effects of fuelwood removal on the diversity of some cavity using birds and mammals in South Africa. *Biological Conservation* 74:77-82.

Eltron, C. S. (1966). *Dying and deadwood,* In: the patterns of animal communities, 217-305. John Wiley & Sons, New York.

Fager, E. W. (1968). The community of invertebrates in decaying oak wood. *Journal of animal Ecology* 37:121-142.

Farrell, B. D., Miller, C. & Futuyma, J. (1992). Diversification at the insect-plant interface. *BioScience* 42(1):34-42.

Furness, C. K. (1979). Some aspects of fuelwood usage and consumption in African rural and urban areas in Zimbabwe. *South African Forestry Journal* 117:10-12.

Franklin, J. F. Shurgat, H. H. & Harmon, K.E (1987). Tree death as an ecological process. *Bioscience* 37(8):550-556.

Wall, J. P. & Reid, N. (1993). Domestic fuelwood use in a rural township in eastern Australia: evidence for resource depletion and implications for management. *Commonwealth forestry Review* 72: 31-37.

Graham, S. A. (1925). The felled tree trunk as an ecological unit. *Ecology* 6(4):397-411.

Gosz, J. R., Likens, G. E. & Borman, F. H. (1973). Nutrient release from decomposing leaf and branch litter in the Hubbard Brook Forest, New Hampshire. *Ecological Monographs* 43:173-191.

Graham, S. A. (1925). The felled tree trunk as an ecological unit. *Ecology* 6(4): 397-411.

Gandar, M. V. (1984). *Firewood in KwaZulu: quantities and consequences.* In: energy for underdeveloped areas. Energy Research Institute, University of Cape Town.

Gashwiler, J. R. (1970). Plant and mammal changes on clear-cut in west central Oregon. *Ecology* 51:1018-1026.

Hamilton, W. D. (1978). *Evolution under bark, In: Mound, L. A. & Waloff, E. (eds) diversity of insects faunas*. Blackwell Scientific Publications. Oxford.

Harcombe, P. A. & Marks, P. L. (1983). Five years of tree death in a Fagus-Magnolia forest, southeast Texas, USA. *Oecologia* 57: 49-64.

Harmon, M. E. (1982). Decomposition of standing dead trees in the southern Appalachian Mountains. *Oecologia* 52:214-215.

Harmon, M. E., Franklin, J. F., Swanson, F. J., Sollins, P. Gregory, S. V., Lattin, J. D., Anderson, N. H., Cline, S. P., Aumen, N. G., Sedell, J. R., Leinkaemper, G. W., Crmack, K. & Cummins, K. W. (1993). Ecology of course woody debris in temperate ecosystems. *Advances in Ecological Research* 15: 133-301.

Hart, S. C. (1999). Nitrogen transformation in fallen tree boles and mineral soil of an old growth forest. *Ecology* 80: 1385-1394.

Hickins, N. E. (1963). *The insect factor in wood decay*. Huntchinson, London.

Hirth, H. F. (1959). Small mammals in old field succession. *Ecology* 40(3):417-425.

Holmes, R. T. & Strges, F. W. (1975). Bird community dynamics and energetic in northern hardwood ecosystems. *Journal of Animal Ecology* 44: 175-200.

Johnston, D. W. & Odum, E. P. (1956). Breeding bird population in relation to plant succession on the piedmont of Georgia. *Ecology* 37(1): 50-62.

Käärik, A. A. (1974). *Decomposition of wood: in: Dickinson, C. H. & Pugh, G. S. F. (eds), biology of plant litter decomposition*. Academic press, New York.

Kavin, K. (2001). Defending deadwood. *Science* 293 (5535): 1579-1581.

Lachat, T., Nagel, P., Cakpo, Y., Attignon, S., Goergen, G., sinsin, B., & Peveling, R. (2006). Deadwood and saproxylic beetle assemblages in a semi-deciduous forest in Southern Benin. *Forest Ecology and Management* 225(1-3): 27-38.

Losey, J. E. & Vaghan, M. (2006). The economic value of ecological services provided by insects. *BioScience* 56 (4): 311-323.

Mannan, R. W., Meslow, E. C. & Weight, H. M. (1980). Use of snags by bird in Douglas-fir forest, Western Oregon. *Journal of Wildlife Management* 44(4): 787-797.

Mainguet, M. (1991). *Desertification, natural background and human mismanagement*. Springer-Verlag, New York.

Maruzane, D. & Cutler, D. (2002). *Firewood in southern Africa with specific reference to woodland management initiative in Zimbabwe*: In: Baijnath & Singh (eds) rebirth of Science in Africa - A shared vision for life and environmental Science, business Print, Pretoria.

Marshall, V. G., Setälä, H. & Trofymow, J. A. (1998). Collembolan succession and stump decomposition in Doglas-fir. *Northwest Science* 72: 84-85.

Mattson, K. G., Swank, W.T. & Waide, J. B. (1987). Decomposition of woody debris in a regenerating clear cut forest in Southern Appalachians. *Canadian Journal of Forest Research* 17: 721-728.

Raphael, M. G. & Morrison, M. L. (1987). Decay and dynamics of snags in the Sierra Nevada, Carlifornia. *Forest Science* 33(3): 774-783.

Raphael, M. G. & White, M. (1984). Use of snags by cavity-nesting birds in the Sierra Nevada. *Wildlife monographs* No 86. 66pp.

Rhoades, F. (1986). Small mammal mycophagy near woody debris accumulations in the Stchekin River Valley, Washington. *Northwest Science* 60(3): 150-153.

Scholtz, C. H. & Holmn, E. (1996). *Insects of southern Africa*. Butterworths, Durban.

Savory, J. G. (1974). Damage to wood caused by microorganisms. *Journal of Applied Bacteriology* 17: 213-218.

Samways, M. J. (1993). Insects in biodiversity conservation: some perspective and directives. *Biodiversity and Conservation* 2: 258-282.

Saniga, M. & Schütz, J. P. (2001). Dynamics of changes in deadwood share in selected beech virgin forests in Slovakia within the development cycle. *Journal of forest Science* 47(12): 557-565.

Sippola, A. L., Sïïtonen, J. & Kallio, R. (1998). Amount and quality of course woody debris in natural and managed coniferous forest near the timberline in finnish Lapland. *Scandinavian Journal of forest Research* 13: 204-214.

Suanders, D. A., Hobbs, R. J. & Margules, C. R. (1991). Biological consequences of ecosystem fragmentation: e review. *Conservation Biology* 5:18-32.

Shackleton, C. M. (1993a). Demography and dynamics of dominant woody species in a communal and protected area of eastern Transvaal Lowveld. *South African Journal of Botany* 59: 569-574.

Shacketon, C. M. (1993b). Fuelwood harvesting and sustainable utilization in a communal grazing land and protected area of the Eastern Transvaal. *Biological conservation* 63: 247-254.

Zar, J. H. (1984). *Biostatistical analysis* (2nd) Prentice Hall, New Jersey.

4

Cell Surface Display

Sharadwata Pan and Michael K. Danquah
Monash University
Australia

1. Introduction

The manipulation of the cell surfaces of prokaryotes (mainly bacteria) and eukaryotes (such as Yeast) has manifested to be an area of stupendous ongoing research, with intelligent widespread applications spanning different arenas of biological sciences (Charbit et al., 1988; Cruz et al., 2000; Francisco et al., 1993; Götz, 1990; Jostock & Dübel, 2005; Keskinkan et al., 2004; Kotrba et al., 1999; Lee & Schnaitman, 1980; Liljeqvist et al., 1997; Martineau et al., 1991; Mizuno et al., 1983; Sousa et al., 1996; Taschner et al., 2002; Wernérus & Ståhl, 2004; Willett et al., 1995; Xu & Lee, 1999). Till date, majority of the surface display systems developed for Gram-negative bacteria involve introducing external peptides into surface-approachable loops of naturally displayed proteins. This sometimes put extreme size restrictions on the displayed components (Wernérus & Ståhl, 2004). However, this problem is more or less resolved since larger proteins could be inserted through some recently developed bacterial display systems for Gram-negative bacteria (Charbit et al., 1988; Cruz et al., 2000; Lee & Schnaitman, 1980; Mizuno et al., 1983; Xu & Lee, 1999). Thanks to some tireless research, it is now evident that the structural properties of the cell wall in Gram-positive bacteria, i.e. the thick peptidoglycan layer, make them suitable candidates for strict laboratory procedures and demanding field applications (Jostock & Dübel, 2005). On the other hand, lower transformation efficiency has been a significant disadvantage of using Gram-positive bacteria (Wernérus & Ståhl, 2004), considering if someone is working with surface-displayed conjunctional libraries for affinity-based selections. However, libraries of significant size could also be obtained for Gram-positive bacteria. Transformation frequencies as high as $10^5 - 10^6$ colony forming units/μg of DNA have been observed for *Staphylococcus carnosus* (Götz, 1990). Until recently, different surface displaying systems have been successfully developed (Lee et al., 2003). Based on their recombinant portfolios, these can be categorized into three principal groups: C-terminal fusion, N-terminal fusion, and Sandwich fusion. Natural occurring surface proteins with distinct restricting signals within their N-terminal part may use a C-terminal fusion mechanism to affix external peptides to the C terminus of that functional portion. In a similar way, a N-terminal fusion system points external proteins to the cell wall by using either *Staphylococcus aureus* protein A, fibronectin binding protein B, *Streptococcus pyogenes* fibrillar M protein, and *Saccharomyces cerevisiae* α-agglutinin, all of which contain C-terminal screening signals. However, in many surface proteins, the whole structure is an essentiality for successful aggregation, primarily because the anchoring regions are absent in their subunits (such as outer membrane proteins or OMPs). Here, the sandwich fusion plays a vital role. *Escherichia coli* PhoE, FimH, FliC, and PapA act as good carriers for sandwich fusion for small peptides (Xu & Lee, 1999).

Exhaustive investigations had been carried out in displaying antigens on the surface of different bacterial species that are not corresponding in structure or evolutionary origin (Charbit et al., 1988; Cruz et al., 2000; Francisco et al., 1993; Götz, 1990; Jostock & Dübel, 2005; Keskinkan et al., 2004; Kotrba et al., 1999; Lee & Schnaitman, 1980; Liljeqvist et al., 1997; Martineau et al., 1991; Mizuno et al., 1983; Sousa et al., 1996; Taschner et al., 2002; Wernérus & Ståhl, 2004; Willett et al., 1995; Xu & Lee, 1999). The motive was to use them as carriers of vaccine-delivery, mainly for immunizations of or relating to mucous membranes. Several mechanisms have been developed to better the activated immunological response by mutual display of adhesins, mainly for targeting to the mucosal epithelium. Today, cheap whole-cell biocatalysts are a reality, thanks to the surface display of some enzymes on genetically engineered bacteria. Another emerging trend is the progressive use of display of metal-binding peptides on bacterial surfaces, resulting in efficient metal-binding capability. These recombinant bacteria may act as biosensors or in the quarantine of heavy metals in specialized bioremediation endeavors. So, it is now possible to synthesize ideal, conceptualized bacteria using these connecting strategies with increased specificity and affinity towards the target metal. This would result in significant usefulness of these types of bioadsorbents (Sousa et al., 1996). Also, a probable way of creating biofilters, biocatalysts or diagnostic devices is by effectual immobilization of these cells on solid supports. A summary of the microbial surface display systems has been done (see Table 1). So cell surface display as a mechanism has been accepted and applied for various biotechnological initiatives encompassing areas as important as vaccine delivery, bioremediation and selection platform (Wernérus & Ståhl, 2004), and an array of recent scientific findings indicate that it will continue to act as a promising tool for applied research in years to come.

2. Concepts and pre-existing surface display approaches

2.1 Surface display in Prokaryotes (gram-positive and gram-negative bacteria)

2.1.1 Gram-negative bacteria

Selection systems for the prokaryotes include cellular and phage display and are based on *E. coli*. This is because of its genetic build-up, culturing and maintenance protocols have been extensively studied and are pretty optimized with assuring reproducibility in laboratory and industrial scale. Majority of the outer membrane of *E. coli* is constituted of proteins, which epitomizes a range of adhering mechanisms for foreign sequences. The basic concept of surface display in gram-negative organisms is shown here (see Fig. 1) and some examples are summarized (see Table 2). Some of the common outer membrane proteins that have been used for surface display are (Jostock & Dübel, 2005):

LamB: LamB gene encodes the outer membrane protein maltoporin of *E. coli* which facilitates the transfer of maltose and maltodextrin across the outer membrane. A large polypeptide library of around 5 million different clones uses Maltoporin as the carrier protein. Metal-identifying polypeptides have been isolated by displaying this library on *E. coli* and selecting on metals such as Gold or Chromium.

OmpT: It is an important member of the Omptin family of proteases that has been surface-displayed in *E. coli*. *E. coli* cells that express effective OmpT could be augmented from cells expressing non-effective OmpT by nearly 5000-fold in a single round by coupling both Fluorescence Activated Cell Sorting (FACS) and Fluorescence Resonance Energy Transfer (FRET). For developing enzymes, the same selection principle has been used.

Lpp-OmpA: It has been used extensively for displaying antigens, antibodies, peptides and enzymes. The Lpp-OmpA system is a combination of Lpp (the first nine N-terminal amino

Carrier protein	Host Organism	Insert size	Fusion	Insert
		Prokaryotes		
Gram-negative				
FimH	*E.coli*	7–52 aa	Intern	Peptide library
Flagellae	*E.coli*	11–302 aa	Intern	Peptide library epitope mapping
Pilin	*E.coli*	7–56 aa	Intern	Peptide epitopes
Intimin	*E.coli*	128 aa	C-terminal	Gene-fragment peptide library
Invasin	*E.coli*	18 aa	C-terminal	Peptide library
LamB	*E.coli*	11-232 aa	Intern	Peptide library
OmpC	*E.coli*	162 aa	Intern	Peptides
PAL	*E.coli*	ca. 250 aa	N-terminal	scFv fragments
Gram-positive				
Protein A	*S. carnosus* / *S. xylosus*	ca. 250 aa	N-terminal N-terminal	scFv fragments
Protein A	*S. carnosus*	Up to 397 aa	N-terminal	Cellulose binding domain
		Eukaryotes		
α-agglutinin receptor	Yeast	Up to 620 aa	C-terminal	MHC Class I and II Cytokines Growth Factors Selectines

Table 1. Some surface display systems in both prokaryotes and eukaryotes that are suitable for the functional screening of molecular aggregation. Here 'aa' symbolizes amino acids. Reproduced from an earlier review (Jostock & Dübel, 2005).

acids and the signal sequence)and an OmpA fragment of the original protein (containing five of the eight membrane covering loops). For displaying on the outer membrane of *E. coli*, heterologous proteins (up to 40 kilodaltons or kDa) can be blended to the C-terminus of the Lpp-OmpA fusion protein. This is also a convenient method for displaying the target antigen.
Inp: A glycosylphosphatidylinositol (GPI)-anchor sequence is responsible for binding the Ice-nucleation protein (Inp) of *Pseudomonas syringae* to the surface of the cell. By this way, it can be used as a carrier (in an effective form) to display enzymes on the surface of *E. coli*. The fact that single-chain antibodies (scFVs) have already been displayed as Inp-fusion proteins on *E. coli* makes this system tailor-made for surface displaying antibody libraries.
Intimin: Adhesins (like Intimin) are expressed by *E. coli* strains (capable for causing diseases in the intestinal tract)on their surfaces. This particularly connects with the destined structures on the host cells (eukaryotic). Coalition partners of up to 128 aa residues, derived from different species, have been practically displayed on *E. coli* K-12 strain surface by displacing the two carboxyterminal domains of the EaeA intimin of *E. coli* O157:H7. A common estimation is that one cell displays around 35 thousand shortened intimin molecules.
FimH: Type 1 fimbriae are a common surface feature of majority of *E. coli* strains. FimH, which can be found on the apex of type 1 fimbriae, helps in binding to sturctures that contain α-D-mannose. Without manipulating the biological function of fimbriae, special sequences (from different species) can be inserted in the C-terminal part of FimH. Scientists have selected

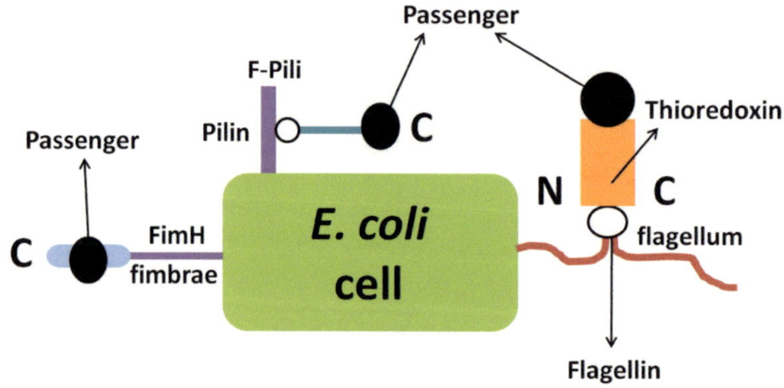

Fig. 1. Cell organelle associated surface display mechanism in Gram negative bacteria. Reproduced and redrawn from (Jostock & Dübel, 2005).

for Ni^{2+} binding clones by constructing an arbitrary peptide library in the FimH protein that can be displayed on *E. coli*.

Invasin: For displaying a peptide library on *E. coli*, a carrier protein in the form of Invasin (of *Y. pseudotuberculosis*), has been in practice. The C-terminal part of invasin binds to integrins and can be displaced by arbitrary peptides (with ten subunits). Peptides with cell binding cabalities may be isolated by library screening on whole mammalian cells.

2.1.2 Gram-positive bacteria

There has been an abundant use of gram positive bacteria in presenting fragments of proteins from different species (between 15 and 459 amino acid residues) (Jostock & Dübel, 2005). However, these applications are cornered to the area of vaccine production, due to the immunological relevance of these gram-positive bacterial strains. Further, a notion of non-trustworthiness prevails in the wider community citing the non-optimization of genetic manipulation of some of these bacteria, as opposite to the scenario in *E. coli*. Genetically altered expression and secretion systems (for proteins) in *Bacillus subtilis* and many other gram positive bacteria are common today (Götz, 1990; Liljeqvist et al., 1997; Wernérus & Ståhl, 2004). The concept of surface display in such organisms is shown here (see Fig. 2) and some examples are summarized (see Table 3).

Till date, the expression of single chain antibodies as fusions to *Staphylococcus aureus* Protein A (SpA) on the non-harmful and food-grade *S. xylosus* and *S. carnosus* strains (Wernérus & Ståhl, 2004) indicates the suitability of Staphylococcal cells as candidates for selecting antibody repositories. However, due to the lower transformation efficiency (as compared to *E. coli*), the use of *Staphylococci* as hosts for conjunctional libraries has suffered. The primary reason being the limitation of the library-size that could be obtained (Wernérus & Ståhl, 2004).

2.2 Surface display in Eukaryotes (Yeast)

The surface display system in yeast demonstrates a C-terminal attachment to the Aga2p subunit of *Saccharomyces cerevisiae* α-agglutinin receptor. This is bound to the Aga1p subunit through two disulphide bonds, which is attached to the β-glucan of the cell wall via covalent bonds. This system has been authenticated for displaying antibody fragments (including Fab fragments), peptides and other protein domains (Jostock & Dübel, 2005). There is large degree

Display system	Displayed protein
Category: Outer Membrane Proteins (OMPs)	
OmpA	Peptides, Malarial antigens
LamB	C3 epitope of poliovirus
	Peptide library
	Peptides
OprF	Malaria epitope
PhoE	Part of FMDV
OmpS	Epitopes
OmpC	$(His)_{162}$
FhuA and BtuB	T7 tag, myc epitope
Lpp'OmpA	Green Fluorescent Protein
	β-lactamase
	PhoA
Invasin	Peptide libraries
EaeA Intimin	Epitope mapping
Inp	CM Cellulose
	Salmobin
	OPH (library)
Category: Autotransporters	
IgA_{β}	CTB, MT
AIDA-I	CTB and peptide antigen
	β-lactamase
Ag43	FimH lectin domain
MisL	Malaria epitope
Other systems	
Peptidoglycan associated lipoprotein	Antibody fragments
TraT	Poliovirus epitope
Pullulanase	β-lactamase

Table 2. Selective examples where Gram-positive bacteria have been used for surface-display applications. Reproduced from an earlier review (Wernérus & Ståhl, 2004).

of similarity between the analysis and selection of yeast displayed libraries to that of bacteria. Healthy, boisterous systems are also a reality (Boder & Wittrup, 1997; Murai et al., 1998; Sousa et al., 1998). The concept of surface display is nicely elaborated in an earlier work (Jostock & Dübel, 2005).

3. Some novel applications of cell surface display technique

Till date, there had been many significant contributions in the area of cell surface display of heterologous proteins. Some of them are categorized into common application areas (see Table 7) and briefly described below, mainly to get an idea of the wide applicability of the surface display technique. Selected examples from an earlier review have been summarized (see Tables 4–6).

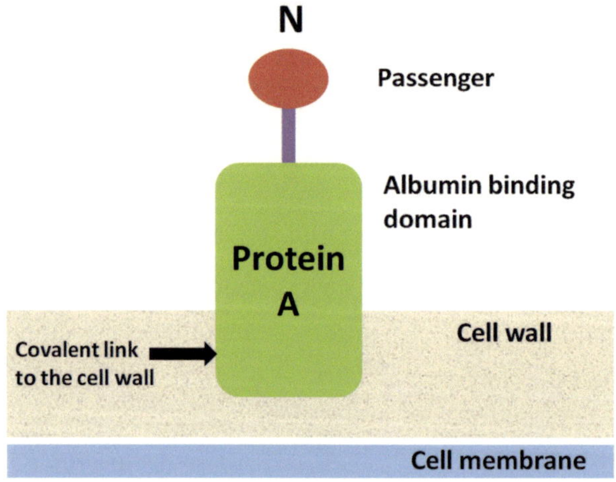

Fig. 2. Cell surface display in Gram positive bacteria. *S. aureus* protein A serves as fusion partner for the surface display. Reproduced and redrawn from (Jostock & Dübel, 2005).

Display system	Displayed protein
Protein A	scFv
	RSV G-protein
	IgA- and IgE-specific affibodies
	Polyhistidyl peptides
	Streptavidin
FnBPB	*Staphylococcus hyicus* lipase, β-lactamase
M6	E7 protein of human papillomavirus
	White-faced-hornet (*Vespula maculata*) antigen
	Tetanus toxin fragment C
	Staphylococcal nuclease
SpaPI	*Bordella pertussis* SI subunit
CwbA	*Yersinia pseudotuberculosis* invasin
CotB	Tetanus toxin
Mtb19	OspA lipoprotein from *Borrelia burgdorferi*
SLH	Tetanus toxin fragment C

Table 3. Selective examples where Gram-positive bacteria have been used for surface-display applications. Reproduced from an earlier review (Wernérus & Ståhl, 2004).

3.1 Vaccine delivery and diagnostic devices

Charbit et al. (1988) demonstrated the expertise of a vector for expressing external polypeptides on the surface of *E.coli*. Their work has formidable potential to create applications. This includes production of an efficient live bacterial vaccine. Liljeqvist et al. (1997) had expressed cholera toxin B subunit (CTB) from *Vibro cholerae* on the surface of two staphylococcal species, *Staphylococcus xylosus* and *Staphylococcus carnosus*. Their work showed enough promise for designing live vaccine delivery systems in bacteria (through the mucosal pathway). According to the authors, further work can be carried out in this area. Rockberg et al. (2008) introduced a remarkable antibody-identified mapping method for epitopes. They expressed antigenic fragments on bacteria and followed it up with antibody-dependent sorting through flow-cytometry. The authors proved that epitope-specific antibodies may be synthesized using bacteria cells. Dou et al. (2009) used surface display technology in bacteria and investigated the pathogenicity of the Japanese Encephalitis Virus (JEV). The authors achieved this by constructing a genetically manipulated *Salmonella typhimurium* BRD509 strain and surface-displayed domain III of the covering protein of the JEV (JEDIII) with the aminoterminal domain of the ice nucleation protein (INPN). They used Western blot and immunohistochemical staining to confirm the surface display. According to the authors, it is now feasible to study the pathogenesis of JEV using their approach. In a recent work, phage display technology has been utilized by Urushibata et al. (2010) to bind antigen-binding (Fab) fragments and single chain variable fragments (scFv) to staphylococcal enterotoxin B (SEB) protein. Their work is noted for developing a unique method for preparing an anti-SEB Fab fragment library. The usefulness of these agents as molecular recognition tools was confirmed by successful application to the SEB determinants from serum by Western blotting. The authors conclude that SEB can be identified by their synthesized scFv and this can even replace anti-SEB immunoglobulins as a cost-effective SEB identification tool.

3.2 Enzymes and biocatalysis

Murai et al. (1998) showed that a yeast cell, which is surface-manipulated with enzymes (alpha-glucosidase and carboxymethylcellulase), acquire the ability to digest cellooligosaccharides. According to the authors, this can be the initiation of the digestion of cellulosic sunstances by *S. cerevisiae* that expresses cellulase genes from different species. As evident from the conclusion of this work, this can be further researched for identifying the next digestion steps. Tsai et al. (2009) through a contemporary work, showed that a single yeast strain containing the required cellulolytic enzymes: two endoglucanases and one exoglucanase (through a displayed minicellulosome) can actively carry out both concurrent and cooperative saccharification and fermentation of cellulose to ethanol. The authors conclude that their overall yield was 0.49 gram of ethanol produced per gram of carbohydrate consumed, which corresponds to 95% of the theoretical value.

3.3 Biosensors and bioadsorbents

Sousa et al. (1996) had displayed poly-His peptides and shown increased adsorption of metals by bacterial cells. From the work, it can be concluded that by expressing poly-His peptides, bacteria may act as adsorbents having metal affinity. Now, it is possible to engineer microorganisms which may facilitate bioadsorption of heavy metal ions. According to the authors, exquisite research opportunities exist for professionals in this particular area. In another interesting work, Sousa et al. (1998) showed that Yeast (CUP1) and mammalian (HMT-1A) metallothioneins can be effectively expressed in *E. coli* as attachments to LamB protein. The authors have clearly demonstrated that these hybrid proteins can be expressed. This has enhanced the natural capability of *E. coli* cells to bind Cd^{2+} ions to about $15-20$ fold.

3.4 Selection platform

Martineau et al. (1991) developed a method to derive and analyze anti-peptide antibodies without actually synthesizing peptides. The peptide of choice was expressed by them as a genetical insert within two separate receiver bacterial proteins (MalE and the LamB proteins from *E. coli*). According to the authors, more work can be done in this frontier. In another

Display system	Organism	Displayed antigen	Animal model	Results
		Gram-negative		
MisL	*S. typhimurium*	Malarial (NANP)	Mice	Ag-specific IgG
LamB	*E. coli*	HbsAg (preS2)	Mice and rabbits (i.v.)	Ag-specific IgG
	E. coli	Polio epitope (C3)	Mice (i.p.)	Ag-specific IgG and IgM
OmpA	*S. typhimurium*	Malarial epitopes (SERP and HRPII)	Mice (orally)	Ag-specific IgG and IgM
Chimaeric OmpA	*S. typhimurium*	Malarial epitope (M3)	Mice (i.p.)	Ag-specific IgG
		Gram-positive		
SpA	*S. xylosus*	RSV antigen	Mice (orally)	Ag-specific IgG
	S. carnosus	Streptococcal protein G/CTB	Mice (i.n.)	Ag-specific IgG and IgM
	S. carnosus	CTB/RSV	Mice (i.n.)	Protection
M6	*S. gordonii*	TTFC	Mice (i.n. and subcut.)	Protection
		LTB and HIV-I epitope V3	Mice (subcut.)	Ag-specific IgG
SpaPI	*S. gordonii*	PTS SI	Mice (i.p.)	Protection
		PTS SI	Mice (orally)	Ag-specific sIgA
SLH	*B. anthracis*	TTFC	Mice (subcut.)	Protection
CotB	B. subtilis	TTFC	Mice (subcut.)	Ag-specific IgG
Lipoprotein Mtb19	*M. bovis*-BCG	OspA from *B. burgdorferi*	Mice (i.n.)	Ag-specific IgG and sIgA

Abbreviations: i.d., intradermal; i.n., intranasally; i.p., intraperitoneally; i.v., intravenously; subcut., subcutaneous; Ag., Antigen; PTS, Pertussis Toxin Subunit

Table 4. Selected examples, where live bacteria with surface displayed antigens have been used as vaccine delivery vehicles. Reproduced from an earlier review (Wernérus & Ståhl, 2004).

Display system	Displayed protein
	Gram-negative
Pullulanase	β-lactamase
Lpp'OmpA	β-lactamase
Inp	*Zymomonas mobilis* levansucrase (LevU)
	Bacillus subtilis CM-cellulose
	Salmobin
AIDA-I	β-lactamase
Inp and Lpp'OmpA	OPH and CBD
	Gram-positive
FnBPB	*S. hyicus* lipase and β-lactamase

Table 5. Selected examples of functionally active enzymes displayed on bacteria. Reproduced from an earlier review (Wernérus & Ståhl, 2004).

wrok, a single chain antibody fragment (scFv), containing the variable heavy and variable light regions from two different monoclonal antibodies had been expressed on the outer

Display system	Displayed protein	Strain
Lpp'OmpA	MT	*E. coli*
LamB	MT (mammalian/yeast)	*E. coli*
LamB	MT (α-domain)	*E. coli*
IgAβ	MT (mouse)	*Pseudomonas putida*
Lpp'OmpA	PC (synthetic)	*E. coli*
Inp	PC (synthetic)	*Maraxella sp.*
SpA	$(His)_6$	*S. carnosus / S. xylosus*
LamB	$(His)_6$	*E. coli*
OmpC	$(His_6)_{12}$	*E. coli*
LamB	HP / CP	*E. coli*
OmpA	HSQKVF	*E. coli*
SpA	Engineered CBD	*S. carnosus*
FimH	Peptide library	*E. coli*

Table 6. Selected examples, where metal-binding peptides and proteins have been expressed on the surface of bacteria for environmental applications. Reproduced from an earlier review (Wernérus & Ståhl, 2004). Here 'MT' stands for Metallothioneins and 'PC' stands for Phytochelatins.

surface of *E. coli* (Francisco et al., 1993). The high level expression of this scFv attachment was shown to bind the hapten with increased compatibility and particularity. Boder & Wittrup (1997) had shown that for manipulating cytokines, antibodies and receptors, display on the cell wall of yeast may be a suitable strategy. However, for effective folding and activity, post translational modification has to be a characteristic of the endoplasmic reticulum. The authors conclude that through this work, kinetic parameters can be distinguished for protein binding to soluble ligands through flow cytometry. Hoischen et al. (2002) showed that external proteins in the cytoplasmic membrane of *E. coli* and *Proteus mirabilis* can be fixed using an ingenious surface display strategy of the membrane. These bacterial strains are steady and lack cell walls. They had fused the reporter protei+n, staphylokinase (Sak) to the membrane-spanning regions of some fundamental membrane proteins from these organisms. The authors confirm that accumulation of the fusion proteins (that are strongly attached to the cytoplasmic membrane) is not a common phenomenon. It is also reported that the protein was confined on the external surface. According to the authors, this technique may generate various application areas which may revolutionize the range of applications of surface display systems. Bessette et al. (2004) demonstrated that it is possible to bind briskly segregated peptides to promptly selected targets with high compatibility. The authors synthesized and screened a large library for binding to some unrelated proteins. These included targets which were previously used in phage display selections like human serum albumin, human C-reactive protein etc. According to the authors, this efficient procedure should be helpful in lot of applications concerning molecular identification since it identifies reagents for peptide affinity. Zahnd et al. (2007) came up with a fascinating gradual procedure to display ribosome selection employing an *E. coli* S30 extract for *in vitro* protein synthesis. The authors agree that in ribosome display, the library range is not restricted by the efficiency of transformation of the bacterial cells. Rather, it is limited by the number of distinct ribosomal complexes that are present in the reaction volume. This dissimilarity is actually the number of ribosomal complexes that show a functional protein. The authors also present a procedure that displays ribosomes through eukaryotic *in vitro* machinery for protein synthesis. Kenrick & Daugherty

Surface displayed proteins	Application area for recombinant bacteria
Antibody fragments	Diagnostic devices
Enzymes	Whole Cell Biocatalysis
Adhesins and antigens	Vaccine delivery
Metal binding peptides	Biosensors and Bioadsorbents
Antibody and peptide libraries	Selection platform

Table 7. Examples of surface displayed proteins and possible application areas for recombinant bacteria (Wernérus & Ståhl, 2004).

(2010) demonstrated an analytical extracting process for affinity maturating ligands with particular given targets. These targets are displayed on the external surface of *E. coli*. By using flow cytometric analysis (involving several parameters), the authors conclude that bacterial surface display proves to be a novel and significant mechanism for the discovery and optimization of peptide ligands that are specific to a particular protein.

4. Concluding remarks

It is now evident that till today, an array of proteins derived from different species have been targeted and expressed on the cell surfaces of Gram-negative or Gram-positive bacteria, and a number of different application areas have been identified. Bacterial surface display will be a continuously growing research area and both Gram-negative and Gram-positive bacteria of various kinds will be thoroughly investigated for different biotechnological applications in the near future. Though several surface display techniques have been developed till date, problems do exist and will continue to haunt researchers. Quality of the peptide library displayed on cell surface and reduced enzyme activity while developing whole-cell biocatalysts are now recognized issues. Another significant challenge is the surface display of multiple proteins or proteins consisting of more than one subunit, which tends to make the cells weak and in some cases, may lead to fatality. However, the ultimate challenge remains the transformation of the numerous laboratory-scale successes in this area to the level industrial productivity. With smarter technologies available, this will happen sooner or later, especially in the areas of bioconversion and peptide library screening. Hopefully, this will pave the way for even more successful commercial applications of cell surface display.

5. References

Bessette, P. H., Rice, J. J. & Daugherty, P. S. (2004). Rapid isolation of high-affinity protein binding peptides using bacterial display, *Protein Engineering, Design and Selection* 17(10): 731–739.

Boder, E. T. & Wittrup, K. D. (1997). Yeast surface display for screening combinatorial polypeptide libraries, *Nature Biotechnology* 15(6): 553–557.

Charbit, A., Molla, A., Saurin, W. & Hofnung, M. (1988). Versatility of a vector for expressing foreign polypeptides at the surface of Gram-negative bacteria, *Gene* 70(1): 181 – 189.

Cruz, N., Borgne, S. L., Hernández-Chávez, G., Gosset, G., Valle, F. & Bolivar, F. (2000). Engineering the Escherichia coli outer membrane protein OmpC for metal bioadsorption, *Biotechnology Letters* 22(7): 623–629.

Dou, J., Daly, J., Yuan, Z., Jing, T. & Solomon, T. (2009). Bacterial cell surface display: A method for studying Japanese Encephalitis Virus pathogenicity, *Japanese Journal of Infectious Diseases* 62(5): 402–408.

Francisco, J. A., Campbell, R., Iverson, B. L. & Georgiou, G. (1993). Production and fluorescence-activated cell sorting of Escherichia coli expressing a functional antibody fragment on the external surface, *PNAS* 90(22): 10444–10448.

Götz, F. (1990). *Staphylococcus carnosus*: A new host organism for gene cloning and protein production, *Journal of Applied Bacteriology Symposium Supplement* (19): 49S–53S.

Hoischen, C., Fritsche, C., Gumpert, J., Westermann, M., Gura, K. & Fahnert, B. (2002). Novel bacterial membrane surface display system using cell wall-less L-forms of Proteus mirabilis and Escherichia coli, *Applied and Environmental Microbiology* 68(2): 525–531.

Jostock, T. & Dübel, S. (2005). Screening of molecular repertoires by microbial surface display, *Combinatorial Chemistry and High Throughput Screening* 8(2): 127–133.

Kenrick, S. A. & Daugherty, P. S. (2010). Bacterial display enables efficient and quantitative peptide affinity maturation, *Protein Engineering, Design and Selection* 23(1): 9–17.

Keskinkan, O., Goksu, M. Z. L., Basibuyuk, M. & Forster, C. F. (2004). Heavy metal adsorption properties of a submerged aquatic plant (*Ceratophyllum demersum*), *Bioresource Technology* 92(2): 197–200.

Kotrba, P., Dolecková, L., Lorenzo, V. D. & Ruml, T. (1999). Enhanced bioaccumulation of heavy metal ions by bacterial cells due to surface display of short metal binding peptides, *Applied and Environmental Microbiology* 65(3): 1092–1098.

Lee, D. R. & Schnaitman, C. A. (1980). Comparison of outer membrane porin proteins produced by Escherichia coli and Salmonella typhimurium, *Journal of Bacteriology* 142(3): 1019–1022.

Lee, S. Y., Choi, J. H. & Xu, Z. (2003). Microbial cell-surface display, *Trends in Biotechnology* 21(1): 45–52.

Liljeqvist, S., Samuelson, P., Hansson, M., Nguyen, T. N., Binz, H. & Ståhl, S. (1997). Surface display of the cholera toxin B subunit on *Staphylococcus xylosus* and *Staphylococcus carnosus*, *Applied and Environmental Microbiology* 63(7): 2481–2488.

Martineau, P., Charbit, A., Leclerc, C., Werts, C., O'Callaghan, D. & Hofnung, M. (1991). A genetic system to elicit and monitor anti-peptide antibodies without peptide synthesis, *Bio/Technology* 9(2): 170–172.

Mizuno, T., Chou, M. Y. & Inouye, M. (1983). A comparative study on the genes for three porins of the Escherichia coli outer membrane. DNA sequence of the osmoregulated ompC gene., *Journal of Biological Chemistry* 258(11): 6932–6940.

Murai, T., Ueda, M., Kawaguchi, T., Arai, M. & Tanaka, A. (1998). Assimilation of cellooligosaccharides by a cell surface-engineered yeast expressing β-glucosidase and carboxymethylcellulase from *Aspergillus aculeatus*, *Applied and Environmental Microbiology* 64(12): 4857–4861.

Rockberg, J., Löfblom, J., Hjelm, B., Uhlén, M. & Ståhl, S. (2008). Epitope mapping of antibodies using bacterial surface display, *Nature Methods* 5(12): 1039–1045.

Sousa, C., Cebolla, A. & Lorenzo, V. D. (1996). Enhanced metalloadsorption of bacterial cells displaying poly-His peptides, *Nature Biotechnology* 14(8): 1017–1020.

Sousa, C., Kotrba, P., Ruml, T., Cebolla, A. & Lorenzo, V. D. (1998). Metalloadsorption by Escherichia coli cells displaying yeast and mammalian metallothioneins anchored to the outer membrane protein LamB, *Journal of Bacteriology* 180(9): 2280–2284.

Taschner, S., Meinke, A., Gabain, A. V. & Boyd, A. P. (2002). Selection of peptide entry motifs by bacterial surface display, *Biochemical Journal* 367(2): 393–402.

Tsai, S. L., Oh, J., Singh, S., Chen, R. & Chen, W. (2009). Functional assembly of minicellulosomes on the Saccharomyces cerevisiae cell surface for cellulose hydrolysis and ethanol production, *Applied and Environmental Microbiology* 75(19): 6087–6093.

Urushibata, Y., Itoh, K., Ohshima, M. & Seto, Y. (2010). Generation of fab fragment-like molecular recognition proteins against staphylococcal enterotoxin B by phage display technology, *Clinical and Vaccine Immunology* 17(11): 1708–1717.

Wernérus, H. & Ståhl, S. (2004). Biotechnological applications for surface-engineered bacteria, *Biotechnology and Applied Biochemistry* 40(3): 209–228.

Willett, W. S., Gillmor, S. A., Perona, J. J., Fletterick, R. J. & Craik, C. S. (1995). Engineered metal regulation of trypsin specificity, *Biochemistry* 34(7): 2172–2180.

Xu, Z. & Lee, S. Y. (1999). Display of polyhistidine peptides on the Escherichia coli cell surface by using outer membrane protein C as an anchoring motif, *Applied and Environmental Microbiology* 65(11): 5142–5147.

Zahnd, C., Amstutz, P. & Plückthun, A. (2007). Ribosome display: Selecting and evolving proteins in vitro that specifically bind to a target, *Nature Methods* 4(3): 269–279.

5

Biological Cr(VI) Reduction: Microbial Diversity, Kinetics and Biotechnological Solutions to Pollution

Evans M. N. Chirwa and Pulane E. Molokwane
University of Pretoria,
South Africa

1. Introduction

The reduction of Cr(VI) to Cr(III) in the environment is beneficial to ecosystems since Cr(VI) is highly toxic and mobile in aquatic systems, whereas Cr(III) is less mobile, readily forms insoluble precipitates and is about 1000 times less toxic than Cr(VI) (Mertz, 1974; NAS, 1974). Similar reactions have been used lately in reducing uranium-6 (U(VI)) to the less mobile tetravalent form U(IV) for possible application in areas around nuclear waste repositories (Chabalala & Chirwa, 2010).

Biological Cr(VI) reduction is limited by its toxicity to the organisms that reduce it. In certain groups of bacteria, the Cr(VI) reduction capability may be transferred across species. Such a possibility was demonstrated in a study by Bopp & Ehrlich (1988) where Cr(VI) reduction genes were transferred on plasmids across different serotypes of *Pseudomonas fluorescens*. In 1992-1993, Wang and Shen (1993) evaluated Cr(VI) reduction activity in a transformed *Escherichia* species formerly known as B1. *E. coli* B1 is metabolically diverse and was demonstrated to function well in a multi-pollutant environment. For example, B1, later designated ATCC 33456, was able to grow on metabolites formed during degradation of aromatic compounds and reduce Cr(VI) to Cr(III) in the process (Chirwa & Wang, 2000). Successful simultaneous removal of Cr(VI) together with organic co-pollutants demonstrated the feasibility of treating pollutants in real-life where Cr(VI) is discharged together with a variety of toxic organic copollutants.

In later years, various isolates of Cr(VI) reducing bacteria have been isolated from different sites around the world showing that the Cr(VI) reducing capability of microorganisms is ubiquitous in nature (Ganguli & Tripathi, 2002; Zakaria *et al.*, 2007, Molokwane *et al.*, 2008). Several organisms have shown adaptability to Cr(VI) exposure by either acquiring resistance to Cr(VI) toxicity or by participating in the detoxification of the environment for their own survival through the conversion of Cr(VI) to the less toxic Cr(III). This chapter evaluates the prospects of application of the biological remediation against Cr(VI) pollution and recent improvements on the fundamental process.

2. Background

Chromium has been used extensively in industrial processes such as leather tanning, electroplating, negative and film making, paints and pigments processing, and wood preservation (Beszedits, 1988). Additionally, chromium has been used as a metallurgical additive in alloys (such as stainless steel) and metal ceramics. Chromium plating has been widely used to give steel a polished silvery mirror coating. The radiant metal is now used in metallurgy to impart corrosion resistance. Its ornamental uses include the production of emerald green (glass) and synthetic rubies. Due to its heat resistant properties, chromium is included in brick moulds and nuclear reactor vessels (Dakiky *et al*, 2002).

Through the above and many other industrial uses, a large amount of chromium (approximately 4,500 kg/d) is discharged into the environment making it the most voluminous metallic pollutant on earth. Almost all chromium inputs to the natural systems originate from human activities. Only 0.001% is attributed to natural geologic processes (Merian, 1984).

Chromium from the anthropogenic sources is discharged into the environment mainly as hexavalent chromium [Cr(VI)]. Cr(VI) – unlike Cr(III) – is a severe contaminant with high solubility and mobility in aquatic systems. Cr(VI) is a known carcinogen classified by the U.S.EPA as a Group A human carcinogen based on its chronic and subchronic effects (Federal Register, 2004). It is for this reason that most remediation efforts target the removal of Cr(VI) primarily.

Chromium is conventionally treated by transforming Cr(VI) to Cr(III) at low pH through the following reduction-oxidation (redox) reaction:

$$Cr_2O_7^{2-} + 14H^+ + 6e^- \rightarrow 2Cr^{3+} + 7H_2O + 1.33v\ (E^0) \qquad (1)$$

(Garrel & Christ, 1965), followed by precipitation as chromium hydroxide ($Cr(OH)_3(s)$) at a higher pH. Because of the difference in electric potential between the two states, substantial amounts of energy are needed to overcome the activation energy for the reduction process to occur. It is therefore assumed that spontaneous reduction of Cr(VI) to Cr(III) never occurs in natural aquatic systems at ambient pH and temperature.

The redox reaction of Cr(VI) to Cr(III) requires the presence of another redox couple to donate the three necessary electrons. Sets of common Cr(VI) reducing couples in natural waters include H_2O/O_2, Mn(II)/Mn(IV), NO_2^-/NO_3^-, Fe(II)/Fe(III), S^{2-}/SO_4^{2-}, and CH_4/CO_2. Compounds such as pyrite (FeS_2) and iron sulphide (FeS) can serve as reducing agents for Cr(VI). Iron sulphide (FeS) is ubiquitous in reducing environments such as saturated soils, sediments, and sludge zones of secondary clarifiers in sewage treatment plants. Cr(VI) reduction by iron sulphides leaves a complex precipitate in solution:

$$Cr(VI)_{(aq)} + 3Fe(II)_{(aq)} \rightarrow Cr(III)_{(aq)} + 3Fe(III)_{(aq)} \qquad (2)$$

$$xCr(III) + (1-x)Fe(III) + 3H_2O \rightarrow (Cr_xFe_{1-x})(OH)_{3(s)} + 3H^+ \qquad (3)$$

where x may vary from 0 to 1 (Eary & Rai, 1988). The precipitate $(Cr_xFe_{1-x})(OH)_3(s)$ is innocuous and unaesthetic, and therefore must be removed from treated water before discharging into the environment. In practice, the removal of byproducts of Cr(VI) reduction such as the Fe-OH complexes may be very difficult and expensive. The final process may require a system operated at low pH ranges (<2.0) for the removal of Fe-OH compounds followed by operation at a much higher pH range (8-9.5) for the removal of the Cr(III) precipitate ($CrOH_3(s)$) (Eary & Rai, 1988).

Chemical treatment can be performed *ex situ* or *in situ*. However, chemical agents to be applied *in situ* must be selected carefully to avoid 'unintended' contamination of the treatment area. The primary problem associated with chemical treatment is the nonspecific nature of the chemical reagents. Oxidizing/reducing agents added to the matrix to treat one metal could transform other metals in the system into mobile and more toxic forms (NAS, 1974). Additionally, the long-term stability of reaction products is of concern since changes in soil and water chemistry might create conditions favoring the remobilization of previously reduced toxic species.

3. Biological Cr(VI) reduction and removal

Microbial Cr(VI) reduction was first reported in the late 1970s when Romanenko and Koren'Kov (1977) observed Cr(VI) reduction capability in *Pseudomonas* species grown under anaerobic conditions. Since then, several researchers have isolated new microorganisms that catalyse Cr(VI) reduction to Cr(V) or Cr(III) under varying conditions (Shen & Wang, 1993; Chirwa & Wang, 1997a; Ackerley *et al.*, 2004; Zakaria *et al.*, 2007). Other researchers have also observed Cr(VI) reduction in consortium cultures isolated from the environment (Chirwa and Wang, 2000; Chen and Gu, 2005). Cr(VI) reduction is shown to be cometabolic (not participating in energy conservation) in certain species of bacteria, but is predominantly dissimilatory/respiratory under anaerobic conditions (Ishibashi *et al.*, 1990). In the latter process, Cr(VI) serves as a terminal electron acceptor in the membrane electron-transport respiratory pathway, a process resulting in energy conservation for growth and cell maintenance (Horitsu *et al.*, 1987).

Most micro-organisms are sensitive to Cr(VI), but some microbial species are resistant and can tolerate high levels of chromate. In bacteria, Cr(VI) resistance is mostly plasmid borne whereas Cr(VI) reductase genes have been found both on plasmids and on the main chromosome. Different resistance strategies have been identified, including:

- extraction of chromate via the transmebrane sulphate shuttle (Brown *et al.*, 2006; Hu *et al.*, 2005);
- counteracting chromate-induced oxidative stress by activating enzymes involved in ROS scavenging (catalase, superoxide dismutase) (Ackerley *et al.*, 2004);
- specialised repair of DNA damage by SOS response enzymes (RecA, RecG, RuvAB) (Hu *et al.*, 2005);
- regulation of iron uptake, which may serve to sequester iron in order to prevent the generation of highly reactive hydroxyl radicals via the Fenton reaction (Brown *et al.*, 2006); and
- extracellular reduction of Cr(VI) to Cr(III) which is then removed easily by reacting with functional groups of bacterial cell surfaces (Ngwenya & Chirwa, 2011).

In a few cases, Cr(VI) resistance has been associated with the regulation of uptake mechanisms such as the sulphate uptake shuttle system. Because of its structural similarity to sulphate (SO_4^{2-}), CrO_4^{2-} in some species crosses the cell membrane via the sulphate transport system (Cervantes *et al.*, 2001). After crossing the membrane, CrO_4^{2-} is reduced to Cr^{3+} which interferes with DNA replication resulting in increased rate of transcription errors in the cell's DNA. Additionally, Cr^{3+} may alter the structure and activity of enzymes by reacting with their carboxyl and thiol groups (Cervantes *et al.*, 2001).

Among the resistance mechanisms listed above, the extracellular reduction of Cr(VI) may be utilised in environmental engineering. Although the process is facilitated by bacteria for their own survival, this process can be used to lower the concentration of Cr(VI) in a contaminated environment.

4. Biological treatment option

Cr(VI) reduction by microorganisms often results in consumption of large amounts of proton as reducing equivalents which results in the elevation of the background pH. The increased pH facilitates the precipitation of the reduced chromium as chromium hydroxide, $Cr(OH)_3(s)$ as shown in Equations 4 & 5 below:

$$CrO_4^{2-} + 8H^+ + 6e^- \xrightarrow{\text{Bacteria} \uparrow 3e^-} Cr^{3+} + 4H_2O \xrightarrow{\text{Neutral pH}} Cr(OH)_3(s) + 3H^+ + H_2O \quad (4)$$

$$3CH_3COO^- + 4HCrO_4^- + 4CrO_4^{2-} + 33H^+ \rightarrow 8Cr^{3+} + 6HCO_3^- + 20H_2O \quad (5)$$

Equation 4 illustrates the general biological Cr(VI) reduction reaction in Cr(VI) reducing bacteria (CRB) reconstructed from redox half reactions whereas Equation 5 illustrates a typical reaction under anaerobic conditions using acetic acid as a carbon source and electron donor. Other fatty acid byproducts of hydrolysis can also serve as electron donors for Cr(VI) reduction (Chirwa & Wang, 2000). The obvious advantage of the above process is that it eliminates the need for addition of chemicals in the precipitation stage of the process.

Several carbon sources and reactor configurations have been evaluated. The performance of microbial cultures in treating Cr(VI) is limited mainly by the toxicity effects and the Cr(VI) reduction capacity of the cells (Shen & Wang, 1994). The latter has been demonstrated in several species of bacteria (Shen *et al.*, 1996; Chirwa & Wang, 2000). The problem of limited Cr(VI) capacity in cells is circumvented by using either continuous-flow or biofilm processes, both of which facilitate continuous replenishment of killed or inactivated cells in the system (Nkhalambayausi-Chirwa & Wang, 2005).

5. Cr(VI) reducing microorganisms

The advent of molecular biology has made possible the identification and characterisation of several Cr(VI) reducing species from the environment. Previously, researchers could only identify microbial species that can be cultured using standard broth and agar media. We soon realise that several species of bacteria are not able to grow on standard culturing and growth media and others depend on complex interrelationships with other organisms in a microbial community. Recently, genetic sequencing of 16S rDNA genes and metagenomic techniques have been used to supplement the conventional methods of species identification and characterisation (Jukes & Cantor, 1969). This allows identification of both culturable and unculturable organisms in environmental samples. It also helps uncover species that have not been identified before. Examples of identified Cr(VI) reducing bacteria and their growth conditions are shown in Table 1.

Table 1 illustrates the whole range of species and growth conditions for Cr(VI) reducing organisms. Most of the bacterial species shown in Table 1 were isolated from chromium (VI) contaminated environments (i.e. sediments, wastewater treatment plants, soil etc). Although earlier isolates grew mostly on aliphatic carbon sources, later studies have shown diversity in the preferred carbon sources and electron donors. For example, some consortium cultures

Name of Species	Isolation Conditions/C-Sources	References
*Achromobacter sp. Str.*Ch1	Anaerobic / Luria Broth; glucose-lactate	Zhu *et al.*, 2008
Agrobacterium radiobacter EPS-916	Aerobic-Anaerobic / glucose-mineral salts medium	Llovera *et al.*, 1993
Bacillus megaterium TKW3	Aerobic / nutrient broth-minimal salt medium-glucose, maltose, and mannitol	Cheung & Gu, 2006
Bacillus sp.	Aerobic/ Vogel-Bonner (VB) broth-citric acid; D-glucose	Chirwa & Wang, 1997a;
Bacillus sp. ES 29	Aerobic / Luria-Bertani (LB) medium	Camargo *et al.*, 2003
Bacillus subtilis	Aerobic / Minimal medium - trisodium citrate and dehydrate glucose	Garbisu *et al.*, 1998
Bacillus drentesis, Bacillus thuringiensis	Aerobic/Luria Betani Broth	Molokwane & Chirwa, 2009
Deinococcus radiodurans R1	Anaerobic / Basal Medium, Lactate, Acetate, Pyruvate, Succinate	Frederickson *et al.*, 2000
Enterobacter cloacae HO1 *strain*	Anaerobic / KSC medium-Sodium acetate	Wang *et al.*, 1989
Escherichia coli ATCC 33456	Aerobic-Anaerobic / Nutrient broth medium; glucose, acetate, propionate, glycerol and glycine	Shen & Wang, 1993
Enterobacter sp	Aerobic/Luria Betani Broth	Molokwane & Chirwa, 2009
Lysinibacilus sphaericus	Aerobic/Luria Betani Broth	Molokwane & Chirwa, 2009
Ochrobactrum sp.	Aerobic / glucose	Zhiguo *et al.*, 2009
Pantoea agglomerans SP1	Anaerobic / acetate	Francis *et al.*, 2000
Pseudomonas fluorescens	Aerobic-Anaerobic / Glucose-Acetate-Pyruvate-Lactate-Succinate	Bopp *et al.*, 1983
Pseudomonas fluorescens LB300	Aerobic / Vogel-Bonner broth	Bopp & Ehrlich, 1988
Pseudomonas putida MK1	Anaerobic / Luria-Bertani -citric acid- Tris-acetic acid	Park *et al.*, 2000
Providencia sp.	Aerobic-Anaerobic / Luria broth (tryptone-yeast extract)	Thacker *et al.*, 2006
Shewanella alga (BrYMT) ATCC 55627	Aerobic-Anaerobic / M9 broth- Glucose	Guha *et al.*, 2001
Shewanella putrefaciens MR-1	Anaerobic / lactate- fumarate	Myers *et al.*, 2000

Table 1. Identified Cr(VI) reducing bacteria.

were shown to grow in the absence of organic carbon sources – utilising only bicarbonate (HCO_3^-) as the carbon source (Molokwane & Chirwa, 2009). The table also illustrates that Cr(VI) reducing microorganisms are ubiquitous in nature. They thrive on a range of carbon sources and are found in almost all possible environments. This in itself shows the feasibility of the biological treatment process as it could be adapted to a wide range of effluent and environmental conditions.

6. Proposed Cr(VI) reduction mechanisms

As stated earlier, Cr(VI) reduction may be cometabolic (not participating in energy conservation) in certain species of bacteria, but could be predominantly dissimilatory/ respiratory under anaerobic conditions in certain species. In the latter process, Cr(VI) serves as a terminal electron acceptor in the membrane electron-transport respiratory pathway, a process resulting in energy conservation for growth and cell maintenance (Horitsu *et al.*, 1987; Ishibashi *et al.*, 1990). In the dissimilatory/respiratory process, electrons are donated from the electron donor to Cr(VI) via NADH (Chirwa & Wang, 1997a).

6.1 Cr(VI) reduction by cytoplasmic enzymes

Although it is proven that specialised Cr(VI) reducing enzymes (reductases) exist inside the Cr(VI) reducing bacterial cells, several components of the cell's protoplasm also reduce Cr(VI). Components such as NADH (NADPH in some species), flavoproteins, and other hemeproteins readily reduce Cr(VI) to Cr(III) (Ackerley *et al.*, 2004). It is therefore expected that the cytoplasm fraction of disrupted cells from most organisms will reduce Cr(VI). Such a reduction process is not metabolically linked but will directly affect the cell since most of the intracellular proteins catalyse a one-electron reduction from Cr(VI) to Cr(V) which also generates harmful reactive-oxygen species (ROS) that cause damage to DNA.

6.2 Cr(VI) reduction by soluble reductase

Of special interest are the Cr(VI) reducing enzymes that are produced deliberately by the cell and exported into the media to reduce Cr(VI). Since protein excretion is an energy intensive process, most of these enzymes are produced constitutively, i.e., they are produced only when Cr(VI) is detected in solution and are therefore highly regulated (Chueng & Gu, 2007). The evidence of extracellular Cr(VI) reduction has been presented by a few researchers using a mass balance of Cr(VI) and its reduced species in media and cells (Shen & Wang, 1993; Chirwa and Wang, 1997b).

In a cellular mass balance evaluation by Chirwa & Wang (1997b), Cr(III) uptake by pelleted cells after centrifugation of a sample of *Pseudomonas fluorescens* LB300 was determined to be only 5% of the initial added Cr as Cr(VI). Cr(III) accumulation in the pelleted cells was determined by measuring the difference in Cr(III) level in solution before and after washing the cells three times in 1.0 N HCl. Results showed that less than 5.0% of Cr(III) was retained in the pelleted cells after 24 hours (0.36 ± 0.04 mg Cr(III)/L in pelleted cells; 8.55 ± 0.22 mg Cr(III)/L in supernatant; 8.62 ± 0.34 mg Cr(III)/L in culture medium). Similar results were obtained earlier by Shen & Wang (1993) with batch cultures of *Escherichia coli* ATCC 33456 in which only about 2.0% of Cr(III) transformed from Cr(VI) remained in the cell pellets.

Extracellular Cr(VI) reduction is beneficial to the organism in that the cell does not require transport mechanisms to carry the chromate and dichromate into the cell and to export the Cr^{3+} into the medium. Both Cr^{6+} and Cr^{3+} react easily with DNA, the presence of which can result in DNA damage and increased rates of mutations. Extracellular reduction of Cr(VI), thus, protects the cell from the DNA damaging effects of Cr(VI). It may be due to this reason that certain species of bacteria have adapted the extracellular Cr(VI) reduction process for survival in Cr(VI) contaminated environments.

From an engineering perspective, using cells that reduce Cr(VI) externally is specifically beneficial since the cells can be separated easily from an expired medium and reused in the reactor system. If Cr(VI) is reduced internally, the resulting Cr(III) will tend to accumulate inside the cell, thus it will be difficult to recover reduced Cr or regenerate the cells.

6.3 Membrane pathway

Microorganisms are known to have evolved biochemical pathways for degrading or transforming toxic compounds from their immediate environment either simply for survival or to derive energy by using the toxic compounds as electron donors or electron sinks. The biotransformation pathways commonly take advantage of the advanced and well conserved membrane electron transport respiratory apparatus within the organisms (Dickerson, 1980). For example, the redox reactions involving some of the metallic pollutants are coupled to the electron transport through electron carriers in the cytoplasmic membrane and the flux of protons through the ATP-synthase. The proton flux and production of ATP through the ATP-synthase generates the required energy equivalents for use in cellular metabolism (Lloyd, 2003).

In other studies, two pathways of Cr(VI) reduction are suggested for gram-negative bacteria (Figure 1). The first mechanism suggests Cr(VI) reduction mediated by a soluble reductase with NADH serving as the electron donor either by necessity (Horitsu *et al.*, 1987) or for maximum activity (Ishibash *et al.*, 1990). The NADH-dehydrogenase pathway is expected to predominate under aerobic conditions. In the second mechanism, Cr(VI) acts as an electron acceptor in a process mediated by a membrane-bound Cr(VI) reductase activity (Horitsu *et al.*, 1987).

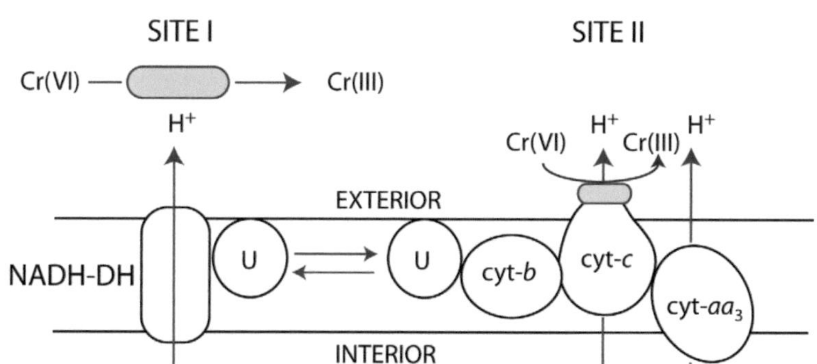

NADH-DH = NADH-dehydrogenase
U = Ubquinone
cyt b = cytochrome *b*
cyt c = cytochrome *c*
cyt aa_3 = cytochrome oxidase

Fig. 1. The conceptual electron transport pathway through the inner cell membrane.

Although the overall reduction of Cr(VI) to Cr(III) ($CrO_4^{2-} \rightarrow Cr^{3+}$) is thermodynamically favorable, this reaction is limited by reaction kinetics under physiological conditions (Garrels & Christ, 1965). The kinetics of Cr(VI) reduction can be improved by coupling Cr(VI) reduction to other energy yielding reactions such as oxidation of organic compounds.

Metabolically linked Cr(VI) reduction associated with the oxidation of NADH was demonstrated in anaerobic cultures of *E. coli* ATCC 33456 under Cr(VI) concentrations below the toxic inhibition threshold (Chirwa & Wang 2000; Nkhalambayausi-Chirwa & Wang, 2001). Under such conditions, Cr(VI) may be used as the principle electron sink and energy is conserved for cell growth and maintenance.
Observations of Cr(VI) reduction under aerobic conditions suggest a cometabolic process where transport of electrons to Cr(VI) does not yield conserved energy for metabolism. In such systems, Cr(VI) is reduced at the expense of metabolic activity in the cells. This was demonstrated using a cumulative mass balance analysis for a continuous-flow biofilm system where cell growth was disrupted during high Cr(VI) loading, but the metabolic activity resumed after Cr(VI) loading was lowered below the toxicity threshold of 10 mg/L (Figure 2). The observed optimum Cr(VI) reduction efficiency just before system

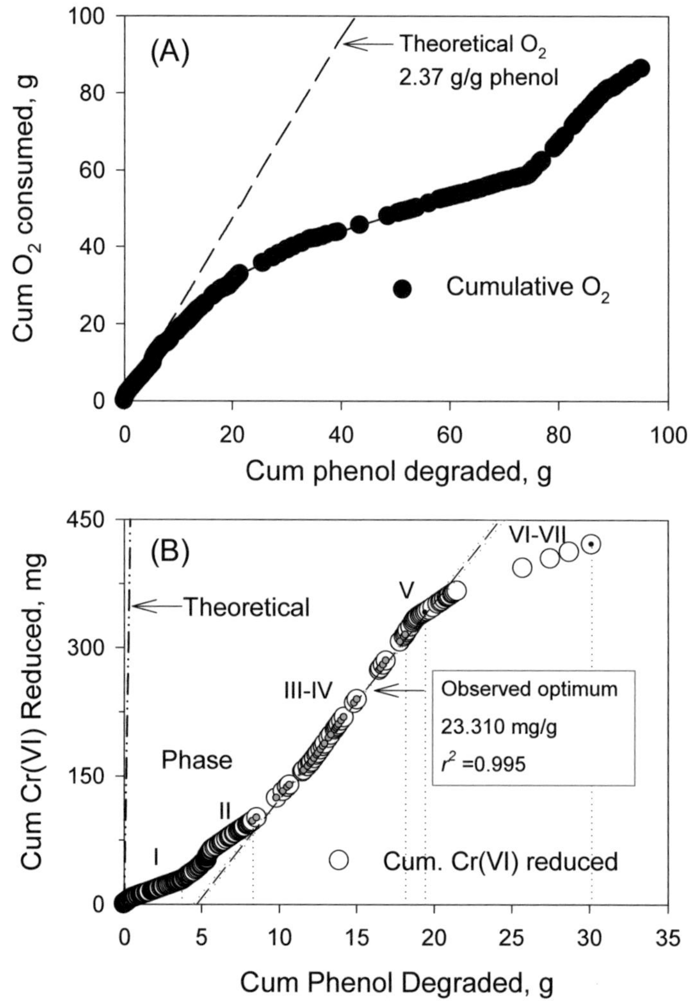

Fig. 2. Cumulative mass balance showing delayed Cr(VI) reduction in a coculture of *Pseudomonas putida* DMP-1 and *Escherichia coli* ATCC 33456 under high loading conditions (Phase III-V). (After Nkhalambayausi-Chirwa and Wang, 2001).

overloading suggested that electrons may be diverted from other biological activities towards Cr(VI) reductase until all Cr(VI) was reduced. If Cr(VI) is still not completely reduced after the cells have sacrificed the maximum number of reducing equivalents to Cr(VI) reduction, then biological activity is completely compromised and the cells may die.

6.4 Genetic regulation

The pioneering work on microbial Cr(VI) reduction was conducted by Romanenko & Koren'Kov (1977) using an unidentified species of *Pseudomonas fluorescens* from Cr(VI) contaminated sediments. Further work revealed that Cr(VI) reduction can either be plasmid borne as was the case with several Pseudomonas species (Bopp and Ehrlich, 1988; Bopp *et al.*, 1983) or located on the chromosomal DNA as is the case with several Bacilli and Enterobacteriaceae (Lu & Krumholz, 2007). Earlier studies also showed that elements carried on the plasmid DNA are transposable across species. This was demonstrated by the creation of *Escherichia coli* ATCC 33456 by transferring the plasmid carrying the Cr(VI) reducing genes from *Pseudomonas fluorescens* LB300 (Shen & Wang, 1993).

So far, only one protein, ChrR, has been demonstrated to receive electrons directly from NADH to achieve Cr(VI) reduction. The protein was purified using classical biochemical techniques from *Pseudomonas putida* (Park *et al.*, 2000) and the resulting homogeneous enzyme successfully catalysed the reduction of chromate. N-terminal and internal amino acid sequence determination of the enzyme allowed the design of appropriate primers to clone the *chrR* gene into *Escherichia coli* (Park *et al.*, 2002). BLAST searching of protein databases with the derived ChrR amino acid sequence revealed a conserved family of proteins whose members are present in a wide range of organisms. Over 40 of these homologs, including the predicted product of a previously uncharacterized open reading frame (*yieF*) from *Escherichia coli*, showed 30% amino acid identity with ChrR. The ChrR and YieF homologs were shown to contain the characteristic signature of the NADH_dh2 family of proteins, which consists of bacterial and eukaryotic NAD(P)H oxidoreductases (Lu & Krumholz, 2007).

The regulation of Cr(VI) reduction in an operon structure was observed in *Bacillus cereus* SJ1 and *Bacillus thuringiensis* strain 97-27 in which the Cr(VI) reduction genes were demonstrated to be upward regulated by the promoter *chrI* which in turn regulated the Cr(VI) resistance gene *chrA1* and arsenic resistance genes *arsR* and *arsB* (He *et al.*, 2010) (Figure 3).

From the observations by He *et al.* (2010) the *chrA1* gene encoding ChrA protein showed the highest amino acid identity (97%) with a homologous protein annotated as chromate transporter in *Bacillus thuringiensis* serovar konkukian str. 97-27. Interestingly, the *chrA1* gene is located downstream of the potential transcriptional regulator gene *chrI*. The region of *chrA1* and *chrI* also contains several putative coding sequences (CDSs) encoding homologs of Tn7-like transposition proteins and a resolvase that is potentially involved in horizontal gene transfer events (Figure 3). ChrI is assumed to control a 26 kb region with a relatively low GC content in *B. thuringiensis* 97-27 (32.8%) which is lower than the average GC content of 35.4% in a corresponding ChrI regulated region in *B. cereus* SJ1.

In both Bacilli, the Chr Operon is interlaced with the arsenic resistance genes including the regulatory genes for the arsenic resistance operon repressor ArsR, arsenic resistance protein

ArsB, arsenate reductase ArsC, arsenic chaperon ArsD and arsenic pump ATPase ArsA (He *et al.*, 2010).

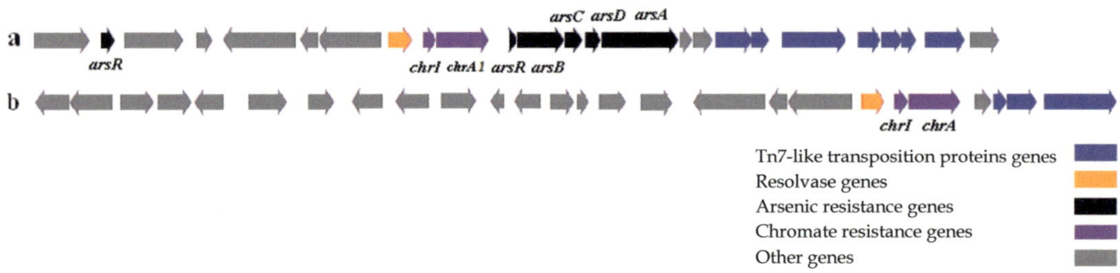

Fig. 3. Comparison of genetic determinants of chromate resistance and chromate reduction between (a) *Bacillus cereus* SJ1 and (b) *Bacillus thuringiensis* serovar konkukian str. 97-27. (After He *et al.*, 2010).

7. Cr(VI) removal from solution - biosorptive processes

In previous studies, it was demonstrated that some species of bacteria possess adsorptive properties that facilitate removal of metal species from aquatic solutions. These adsorptive properties are dependent on the distribution of reactive functional groups on the cell wall surfaces of bacteria, such as; carboxyl, amine, hydroxyl, phosphate and sulfhydryl groups (Parmar *et al.*, 2000). Available information is mostly based on studies conducted under aerobic conditions. There is limited information on microbial adsorptive behaviour under oxygen stressed conditions and toxic environments.

The only available information is on the adsorptive ability of sulphate reducing bacteria for toxic metals including radionuclides (Bruhn *et al.*, 2009). However, as yet, there is lack of knowledge on the nature of the surface reactive groups on SRB cell surfaces that account for its high metal adsorption ability.

In a recent study, Ngwenya & Chirwa (2011) investigated the chemical nature of the cell surfaces of a sulphate reducing bacteria consortium and its interaction with mono- and divalent cations under anaerobic conditions. The study utilized a surface complexation modelling approach to predict the trends of the adsorption of the cationic species.

In the above study, the distribution of functional groups and adsorption reactions on SRB cell surfaces were characterised using a combination of Gram potentiometric titrations, FTIR, and surface complexation modelling. Four types of binding sites were identified: site 1 corresponding to carboxylic acid functional groups (pK_a = 4-5); the near-neutral site 2 corresponding to phosphates (pK_a =6-7), and sites 3 and 4 corresponding to basic sites and phenolic sites (pK_a = 8-12). The most abundant proton binding sites belonged to site 4 (hydroxyl/amine group) and accounted for about 40% of the total concentration of binding sites for the consortium. The effect of ionic strength was also evident from the metal ion adsorption studies. A decrease in metal adsorption was observed at higher ionic strengths. These results promise feasibility of application for recovery of adsorbed metallic species for reuse and regeneration of the cells.

Since the bacteria cell walls show an adsorptive capacity for cationic species, reduction of the oxyanionic chromate (CrO_4^{2-}) to Cr(III), which exists in solution at lower pH as Cr^{3+}, could be necessary for effective removal Cr(VI) from solution. In spite of the effectiveness of biosorption in removing Cr(VI), past studies have strongly supported precipitation as the primary removal mechanism of reduced chromium (Shen & Wang, 1993; Chirwa & Wang, 1997b).

8. Biofilm systems

Microorganisms in nature and in reactor systems rarely grow as separate cells. The microorganisms form complex communities either in the form of agglomerations called flocs or as biofilm on the surfaces of inanimate objects and other organisms. The performance of a microbial culture is not only a function of its capability to degrade or transform a pollutant but also the configuration of the community in which it resides. There are complex interrelationships that occur within the microstructure that affect the availability of substrates, symbiotic existence through toxicity shielding of more sustainable species, and transfer of metabolites to organisms that could otherwise not grow on the only available primary substrate in the bulk liquid.

Nkhalambayausi-Chirwa and Wang in 2001 took advantage of the complex structure of the micriobial biofilm to improve the performance of *Escherichia coli* ATCC 33456 in reducing Cr(VI). The microorganisms in the biofilm could benefit from the spatial and physiological heterogeneity within the biofilm community (Stoodley *et al.*, 1999). In this case, phenol degrading species and Cr(VI) reducing species were grown together in a biofilm reactor such that *E. coli* utilised the anaerobic conditions in deeper layers of the biofilm for growth and Cr(VI) reduction whereas *P. putida* degraded the primary carbon source (phenol) into organic acid metabolites (Nkhalambayausi-Chirwa & Wang, 2001). In so doing, *P. putida* detoxified the environment for *E. coli* and provided secondary carbon and energy sources for *E. coli*.

The operational model of the biofilm system is shown in Figure 4 with the dissolved species represented as C_B, P_B, and U_B in the bulk liquid and $C_{(t,x)}$, ($P_{(t,x)}$), and $U_{(t,x)}$ in the biofilm zone, where C = Cr(VI) concentration (mg/L), P = phenol concentration (mg/L), and U = metabolites concentration (mg/L). The generic representation of dissolved species concentration in the biofilm is given by, $\mathbf{y}_{(t,x)}$. The particulate matter in the reactor consisted of *P. putida* (X_P), *E. coli* (X_E), and inert biomass (X_I). The subscript B in the biomass terms indicates unattached biomass in the bulk liquid.

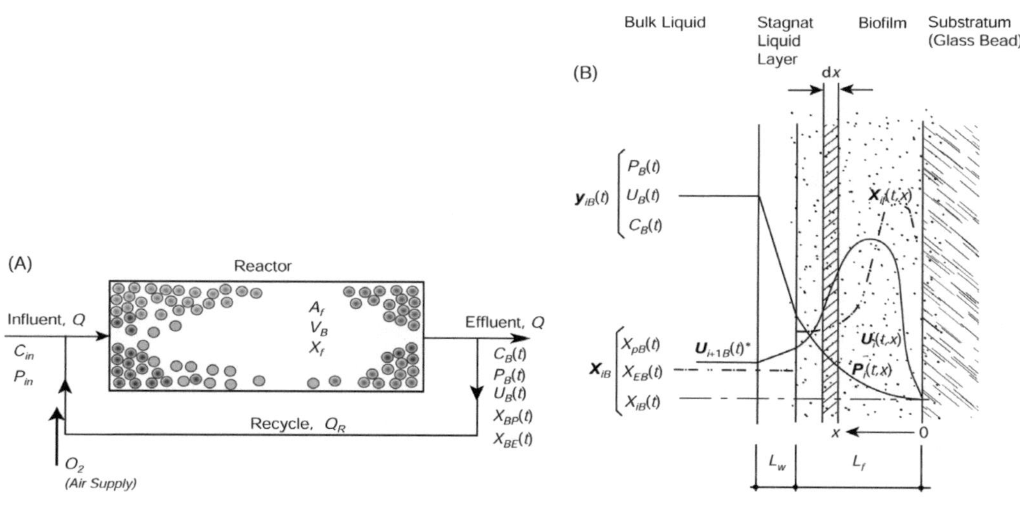

Fig. 4. Conceptual mixed-culture biofilm model for (a) control volume space, and (b) biofilm environment.

In the above model, the primary carbon source (phenol), Cr(VI), and O_2 diffuse into the biofilm where they are taken up by living organisms. A stagnant liquid layer of thickness L_w inherently resists the transport of the dissolved species into the biofilm resulting in generation of a concentration gradient towards the liquid/biofilm interface. Since phenol is toxic to the *E. coli* species used in the study, *P. putida* out-competed *E. coli* in the outer layers of the biofilm. And since *P. putida* is obligately anaerobic, it was out-competed by *E. coli* as O_2 became scarce deeper in the biofilm.

The above system only illustrates the complex nature of the interdependent systems in a near natural environment. Some of these processes can be engineered but some may be lost during the implementation of bioremediation process. During the process of developing the above described coculture, many pairs of aromatic compound degrading species and Cr(VI) reducing species of bacteria were tested but, in most cases, one of the species could be out competed due to susceptibility to toxicity or slow growth.

9. Diffusion/Reaction model

The removal of the dissolved species and cell growth is represented by a set of diffusion-reaction partial differential equations (PDEs) with the conversion reactions occurring mainly inside the biofilm. The PDEs represent a mass balance across an infinitesimal biofilm section (δz) parallel to the substratum surface (Figure 4b) as follows:

$$\frac{\partial(\hat{\mathbf{u}})}{\partial t} = \varepsilon(t)\cdot\frac{\partial(\mathbf{j}_{\hat{\mathbf{u}}})}{\partial z} + \mathbf{r}_{\hat{\mathbf{u}}f} \tag{6}$$

$$\frac{\partial(\hat{\mathbf{x}})}{\partial t} = \varepsilon(t)\cdot\frac{\partial(\mathbf{j}_{\hat{\mathbf{x}}})}{\partial z} + \mathbf{r}_{\hat{\mathbf{x}}f} \tag{7}$$

where: $\mathbf{j}_{\hat{\mathbf{x}}} = D_{w\hat{\mathbf{x}}}\,\partial(\hat{\mathbf{x}})/\partial z$, mass flux rate of biomass ($ML^{-2}T^{-1}$), $\mathbf{r}_{\hat{\mathbf{u}}f}$ = the vector of removal rates of dissolved species in the biofilm ($ML^{-3}T^{-1}$), $\mathbf{r}_{\hat{\mathbf{x}}f}$ = the vector of biomass production rates in the biofilm zone ($ML^{-3}T^{-1}$), and ε = is a biofilm porosity constant (V_{fvoids}/V_{ftotal}). The movement of cells across the biofilm is induced by physical displacement due to growth whereas dissolved species are transported by diffusion. Thus, the values of the j terms for cells are expected to be lower (by orders of magnitude) than the j terms for dissolved species in the biofilm.

The outer and inner boundary conditions for dissolved species $\hat{\mathbf{u}}$ and biomass $\hat{\mathbf{x}}$ are defined by:

$$\mathbf{j}_{\hat{\mathbf{u}}} = k_{L\hat{\mathbf{u}}}\cdot\left(\hat{\mathbf{u}}_B(t) - \hat{\mathbf{u}}_{fs}(t,L_f)\right),\quad z = L_f,\ \text{outer boundary} \tag{8}$$

$$\mathbf{j}_{\hat{\mathbf{x}}} = \lambda(u)\cdot\hat{\mathbf{x}}_f\cdot L_f,\quad z = L_f,\ \text{outer boundary} \tag{9}$$

$$\mathbf{j}_{\hat{\mathbf{u}}} = 0,\quad z = 0,\ \text{inner boundary} \tag{10}$$

$$\mathbf{j}_{\hat{\mathbf{x}}} = 0,\quad z = 0,\ \text{inner boundary} \tag{11}$$

where $k_{L\hat{u}} = D_{w\hat{u}}/L_w$, is the mass transfer rate coefficient (L^2T^{-1}), and $\hat{u}_{fs}$ = dissolved species concentration at the liquid/biofilm interface (ML^{-3}).

The above equations were simulated successfully using optimised reaction rate parameters from batch studies and dynamic parameters from the continuous flow biofilm reactor systems (Nkhalambayausi-Chirwa & Wang, 2005) (Figure 5). The dynamic parameters were estimated from the data obtained from the operation of the reactor at 24 hours hydraulic retention time (HRT) (Phase I-VI). The rest of the phases (VII-XVIII) were simulated using the optimised parameters. The results showed a high predictive accuracy as the model accurately tracked the trends in effluent concentrations for both the electron donor (phenol) and the electron sink (Cr^{6+}).

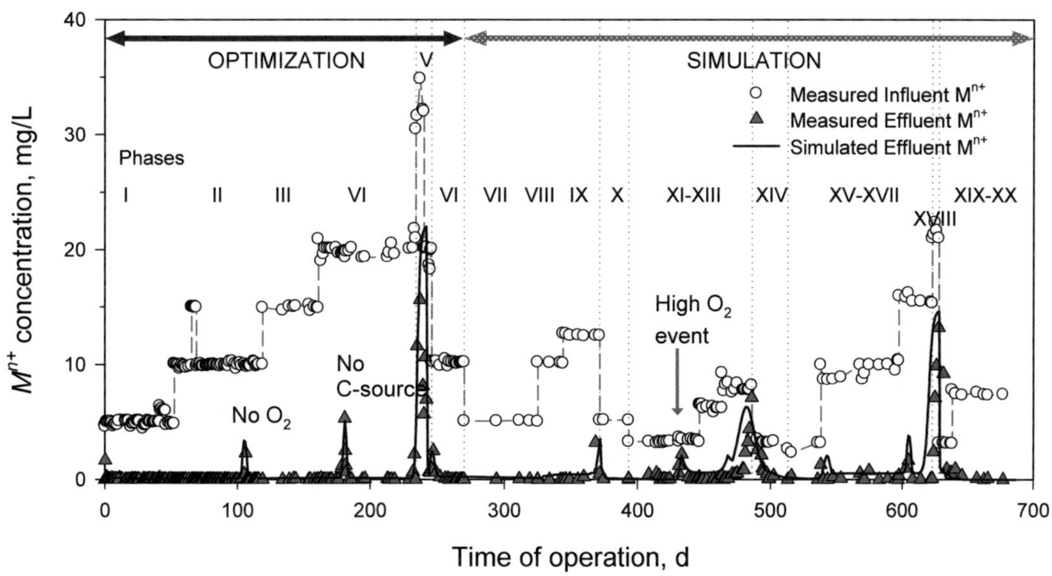

Fig. 5. Simulation of Cr(VI) (M^{6+}) removal in a coculture biofilm system under different HRTs: 24 h (Phase I-VI); 11.7 h (Phase VII-X); 6 h (Phase XI-XIV); 17.9 h (Phase XV-XVIII).

10. *In situ* barrier systems

Several types of treatment walls have been tested in the attenuation of the movement of metals in groundwater. Trench materials have been investigated including zeolite, hydroxyapatite, elemental iron, and limestone (Vidic & Pohland, 1996). Elemental iron has been tested for chromium (VI) reduction and other inorganic contaminants (Powell *et al.*, 1995) and limestone for lead precipitation and adsorption (Evanko & Dzombak, 1997). Biological Permeable Reactive Barriers (BPRBs) use microorganisms as reactants rather than chemical reactants to removal pollutants. Specific application of BPRBs removal of Cr(VI) in groundwater has not been attempted. This has been both due to the unavailability of microorganisms capable of growing under the nutrient deficient groundwater conditions and lack of information on the fate of the reduced chromium species in the barrier.

Recently, the group at University of Pretoria has evaluated the Cr(VI) reduction performance of several anaerobic species of bacteria in microcosm systems simulating groundwater conditions (Molokwane & Chirwa, 2009). The microorganisms were isolated locally to avoid the dilemma of using imported bacteria which is difficult to get and in most instances not allowed by law.

10.1 *In situ* barrier concept

Permeable reactive barriers are an emerging alternative to traditional pump-and-treat systems for groundwater remediation. Such barriers are typically constructed from highly impermeable emplacements of materials such as grouts, slurries, or sheet pilings to form a subsurface "wall." Permeable reactive barriers are created by intercepting a plume of contaminated groundwater with a permeable reactive material (Figure 6). For physical chemical processes such as described above, the reactive materials need to be replenished or replaced after a certain time of operation, a process which is extremely expensive and in some cases not practical. Using microorganisms as the main reactants aims at achieving a self-replenishing system since the bacteria can regenerate themselves. For biodegradable compounds that can be mineralized to CO_2 and H_2O such as petrochemical pollutants, this works perfectly. Unfortunately, metals can only be converted from one form to another such that the converted form may be trapped in the barrier material until measures are taken to remobilise the pollutant to clean the barrier.

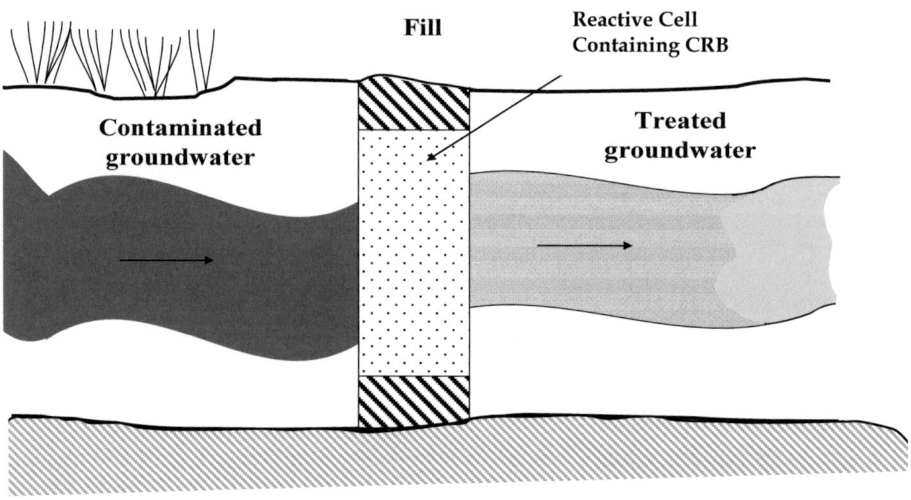

Fig. 6. Elevated view of a permeable reactive barrier configuration for groundwater treatment.

10.2 Application to Cr(VI) and toxic metal removal

As stated earlier, no full scale applications of BPRBs for treating Cr(VI) have been attempted thus far. The group at the University of Pretoria has been evaluating several remediation scenarios for *in situ* treatment of Cr(VI) in groundwater environments. One possibility is *in situ* bioinoculation in which the Cr(VI) reducing mixed-culture of bacteria is injected into the aquifer and allowed to acclimate to the new conditions. This facilitates *in situ* selection for adaptable organisms. In order for the organisms to flourish in the new environment, the prevailing conditions in the environment must just be suitable for the organisms and this is difficult to predict in advance.

A more futuristic approach is the *in situ* molecular augmentation in which transposable elements carrying the metal reducing genes could be introduced into the environment to be taken up by native bacteria in the environment. Upon assimilation of the foreign genetic elements, the native bacteria could then become competent in neutralising the targeted pollutant(s). In this way, importation of foreign bacteria across ecosystems could be avoided. Genetic carriers such as transposons and plasmids have been used in the experiments to evaluate this process by shuttling genetic information for toxic metal remediation into native species that are already best suited to the target environment. Several species of bacteria are capable of picking up and retaining circular fragments of DNA called Broad-Host-Range Plasmids which may be engineered to carry specific genes for the degradation of xenobiotic compounds and transformation of toxic metals (Vincze and Bowra, 2006).

A similar process can be applied using genetically engineered linear DNA called transposons. Although studies have been conducted using these techniques in laboratory microcosms, the application in actual environments has not been attempted (Hill *et al.*, 1994). In the future, it is foreseeable that these methods will find wide application for the new varieties of recalcitrant pollutants being discharged into the environment from several sources.

10.3 Microcosm performance

Cores from an actual contaminated site were set up in the laboratory as microcosm reactors as shown in Figure 7. Contaminant loading was simulated by gravity feeding as is the case in open aquifers a representative Cr(VI) polluted site in Brits (North West Province, South Africa). The experimental systems were installed and operated as packed-bed reactors. All microcosm reactors were operated under a feed concentration of 40 mg/L, representing the observed concentration at the actual site (Brits). 1 mL samples drawn from the influent and effluent were centrifuged at 6000 rpm (2820 × g) for 10 minutes to remove soil particles followed by analysis for Cr(VI) and total Cr as described below.

The microcosm reactors were operated without any added organic carbon sources in the feed solution and no minerals apart from those already found in the soil. Since the system was being developed for possible application in the groundwater environment, introduction of potentially polluting organic carbon sources is not desirable. Autotrophic organisms in the soil are thus expected to use bicarbonate (HCO_3^-) as carbon source and nutrients from soil and decaying vegetation overlying the soil. Efforts are under way to characterise the composition of the organic matter coming from the soil using TOC, DIC, and GC/MS analysis.

The experiments consisted of two non-sterile reactors (R1 and R4) containing native bacteria from the soil, two sterile reactors (R2 and R5) sterilized by autoclaving at 121°C for 30 min., and two consortium inoculated non-sterile reactors (R3 and R6) containing bacteria from dried activated sludge and native soil bacteria. All reactors were operated under a feed concentration of 40 mg/L.

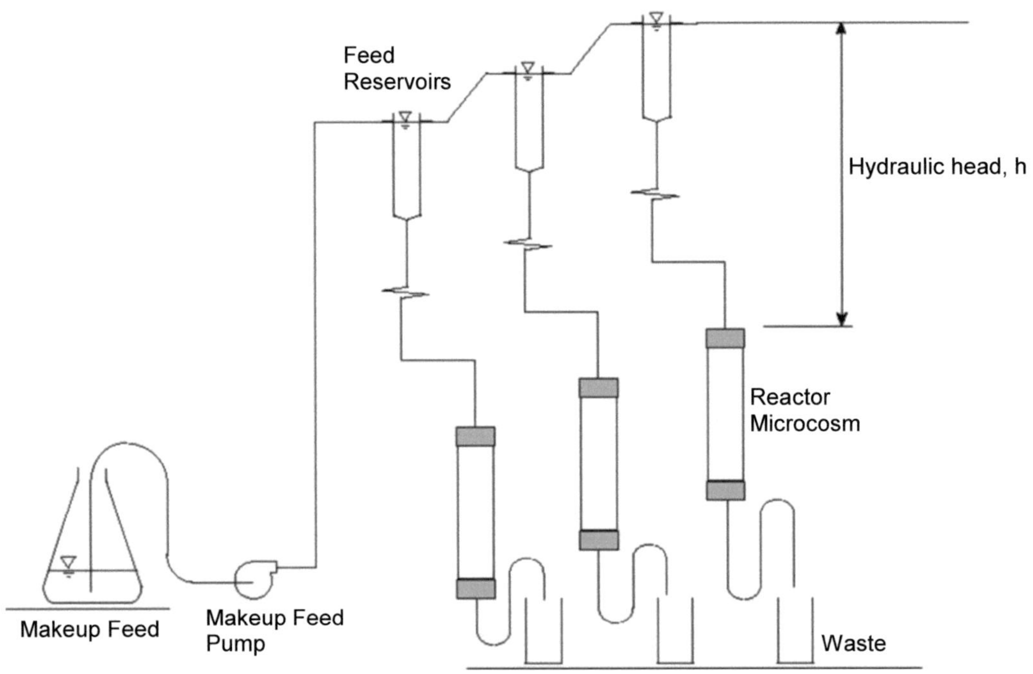

Fig. 7. Experimental setup of the gravity-fed microcosm reactor system.

Columns that experienced severe short-circuiting (R4 and R5) were discontinued. Only reactors R1, R2, R3, and R6 were fully tested. Since the cores were extracted from approximately the same depth at the site, the resistance to flow was almost the same with higher flow rates observed in Reactors 1 and 3.

Data collected showed that one of the columns inoculated with Cr(VI) reducing bacteria (R6) achieved near complete removal of Cr(VI), however, the effectiveness of removal was relatively low at a higher hydraulic loading rate (data not shown). Chromium removal of approximately 95% was observed in the slow feeding reactor R6 (flow rate, Q = 0.310 cm^3/hr) (Figure 8). The removal rate was lower, approximately 80%, in the column with a higher flow rate of 0.608 cm^3/hr (R3). No Cr(VI) removal was observed in the sterilised and in the non-inoculated (native bacteria) controls. The performance of the reactors under different loading conditions is summarised in Table 2.

These experiments clearly show that it is possible to introduce microbial cultures into the environment in a controlled way to achieve Cr(VI) reduction in flowing water. The results do not show how the reduced Cr species, suspected to be predominantly Cr^{3+}, could be remobilised and extracted from the barrier zone once it starts affecting the hydraulic conductivity of the barrier.

10.4 Microbial culture analysis

10.4.1 Characteristics of initial consortium

The robustness of the barrier system was evaluated by monitoring the survival of microorganisms from the Cr(VI) reducing inoculum in the microcosm simulating the aquifer environment. The original inoculum was obtained from dry sludge from sand drying beds

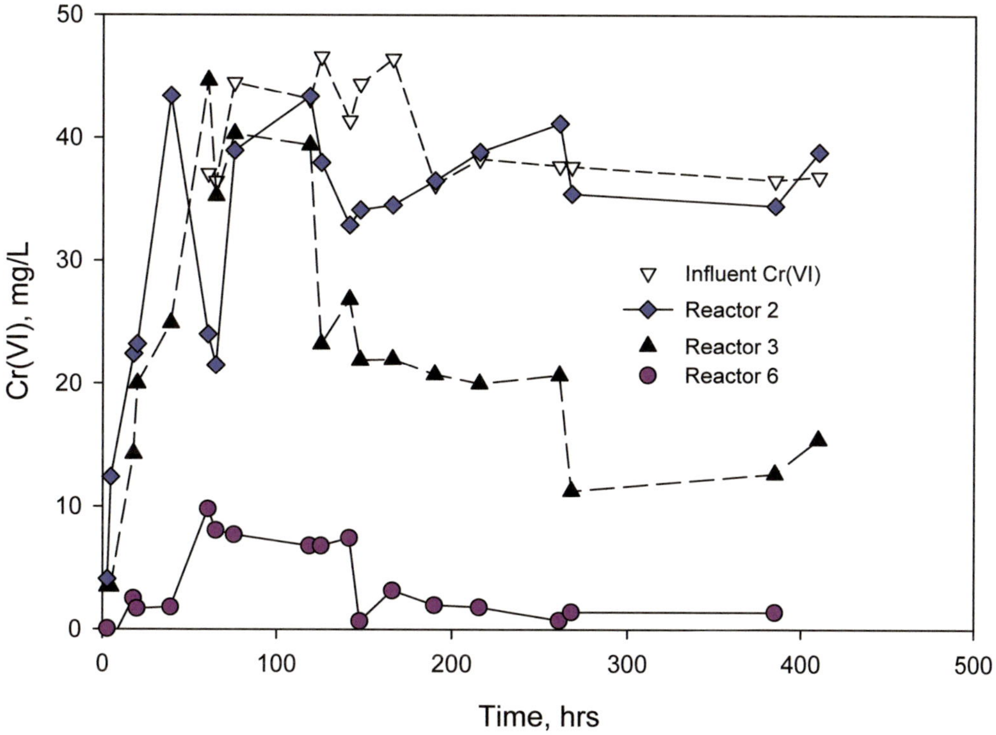

Fig. 8. Cr(VI) reduction in microcosm reactors: Reactor 2 = sterilised column, Reactor 3 = inoculated non-sterile reactor, and Reactor 6 = inoculated non-sterile reactor.

at Brits Sewerage Works (Brits, SA). The sludge bacteria used to inoculate enrichment cultures under microaerobic conditions (under 100 mg/L Cr(VI)) and the colonies isolated based on morphology were further purified and analysed. The predominant species under these enrichment conditions were the Gram-positive *Bacilli* mainly due to inhibition of anaerobic species by oxygen in the sample. Partial sequences of 16S rRNA matched the *Bacillus* groups - *Bacillus cereus* ATCC 10987, *Bacillus cereus* 213 16S, *Bacillus thuringiensis* (serovar finitimus), *Bacillus mycoides* - and two Microbacterium group - *Microbacterium foliorum* and *Microbacterium sp.* S15-M4 (Table 3). A phylogenetic tree was constructed for the species from purified cultures grown under aerobic conditions based on a basic BLAST search of rRNA sequences in the NCBI database (Figure 9).

Reactor Number and Type	**Effluent Cr(VI) Conc. mg/l**	**Effluent Cr(III) Conc. mg/l**	**Cr(VI) Removal %**
Native-soil R1	39.0 ± 2.0	0.0 ± 0.0	0.0 ± 0.0
Non inoculated R2	37.8 ± 1.5	0.0 ± 0.0	0.0 ± 0.0
Inoculated R3	6.7 ± 0.8	1.5 ± 0.4	80 ± 3.6
Inoculated R6	1.9 ± 0.3	3.2 ± 1.1	95.3 ± 1.4

Table 2. Performance of gravity-fed microcosm reactors operated under an influent Cr(VI) concentration of 40 mg/L (0.310 cm^3/h).

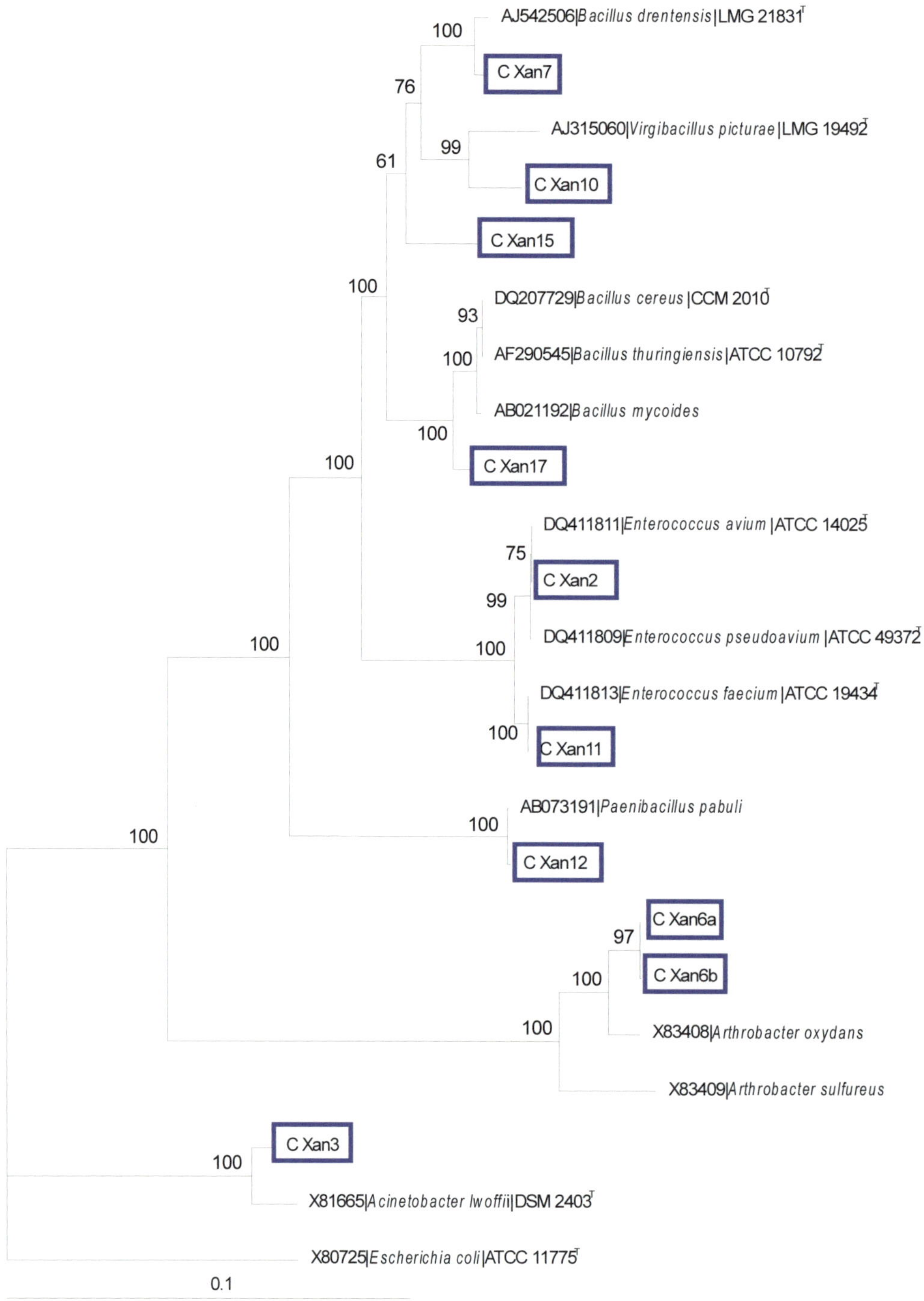

Fig. 9. Phylogenetic tree of species from Brits dry sludge reflecting microbial diversity under anaerobic conditions.

Inoculation culture Consortium		Culture in Reactor 3 and 6 at end of experiment	
Type	**Predominant species**	**Type**	**Predominant species**
X1	*Bacillus cereus* 213 16S, *Bacillus thuringiensis* 16S	A	*Pantoea* or *Enterobacter sp.*
X2	*Bacillus cereus* ATCC 10987, *Bacillus thuringiensis* str. Al Hakam	B	*Bacillus sp.* possibly *Bacillus thuringiensis/ cereus* group
X3	*Bacillus cereus* ATCC 10987, *Bacillus thuringiensis* str. Al Hakam	C	*Pantoea* or *Enterobacter sp.*
X4	*Bacillus mycoides* BGSC 6A13 16S. *Bacillus thuringiensis serovar finitimus* BGSC 4B2 16S	D	*Lysinibacillus sphaericus* strain BG-B111, *Bacillus sp.* G1DM-64, *Bacillus sphaericus*
X5	*Bacillus mycoides* BGSC 6A13 16S. *Bacillus thuringiensis serovar finitimus* BGSC 4B2 16S	E	*Bacillus sp.* possibly *Bacillus thuringiensis/ cereus* group
X6	*Bacillus mycoides* BGSC 6A13 16S. *Bacillus thuringiensis serovar finitimus* BGSC 4B2 16S	F	*Bacillus sp.* possibly *Bacillus thuringiensis/ cereus* group
X7	*Bacillus mycoide* BGSC 6A13 16S. *Bacillus thuringiensis serovar finitimus* BGSC 4B2 16S	G	*Bacillus cereus* strain ZB

Table 3. Microbial culture changes after operation of the microcosms reactors for 15 days under an influent Cr(VI) concentration of 40 mg/L.

10.4.2 Characterisation of microcosm bacteria (after 15 days)

After operating the reactors under oxygen stressed conditions in the presence of other soil bacteria, a community shift was expected. In reactors R3 and R6, the soil contained a wide range of soil dwelling species of bacteria as well as the newly introduced bacteria from the sand drying bed sludge. The microbial dynamics monitored by the 16S rRNA fingerprinting showed a decrease in culturable species after exposure to Cr(VI) as shown in Tables 3. Only the *B. cereus* and *B. thirungiensis* serotypes persisted either due to resilience against toxicity or adaptation to the changing conditions in the reactor. The *Lysinibacillus* group is also a well known sludge bacteria. Both Bacilli (*B. cereus* and *B. thuringiensis*) and the *Lysinibacillus* species contain well known Cr(VI) reducing serotypes such as *Bacillus* K1 (Shen *et al.*, 1996), *Bacillus cereus, Bacillus thirungiensis* (Camargo *et al.*, 2003), and *Lysinibacillus sphaericus* AND 303 (Pal *et al.*, 2005).

10.4.3 Culture composition at the end of experiment

Several species from the original sludge culture disappeared from the consortium after operating the microcosm reactors for 17-20 days. Instead, other species not originally observed appeared in the reactors (Figure 10). Some species in the samples also showed associations with gram-negative species belonging to the Enterococcus and Escherichia groups. These results confirmed the adaptability of the cultures at the community level. Linked with the performance data, the results suggest that more competent species were selected after a long time of exposure to Cr(VI).

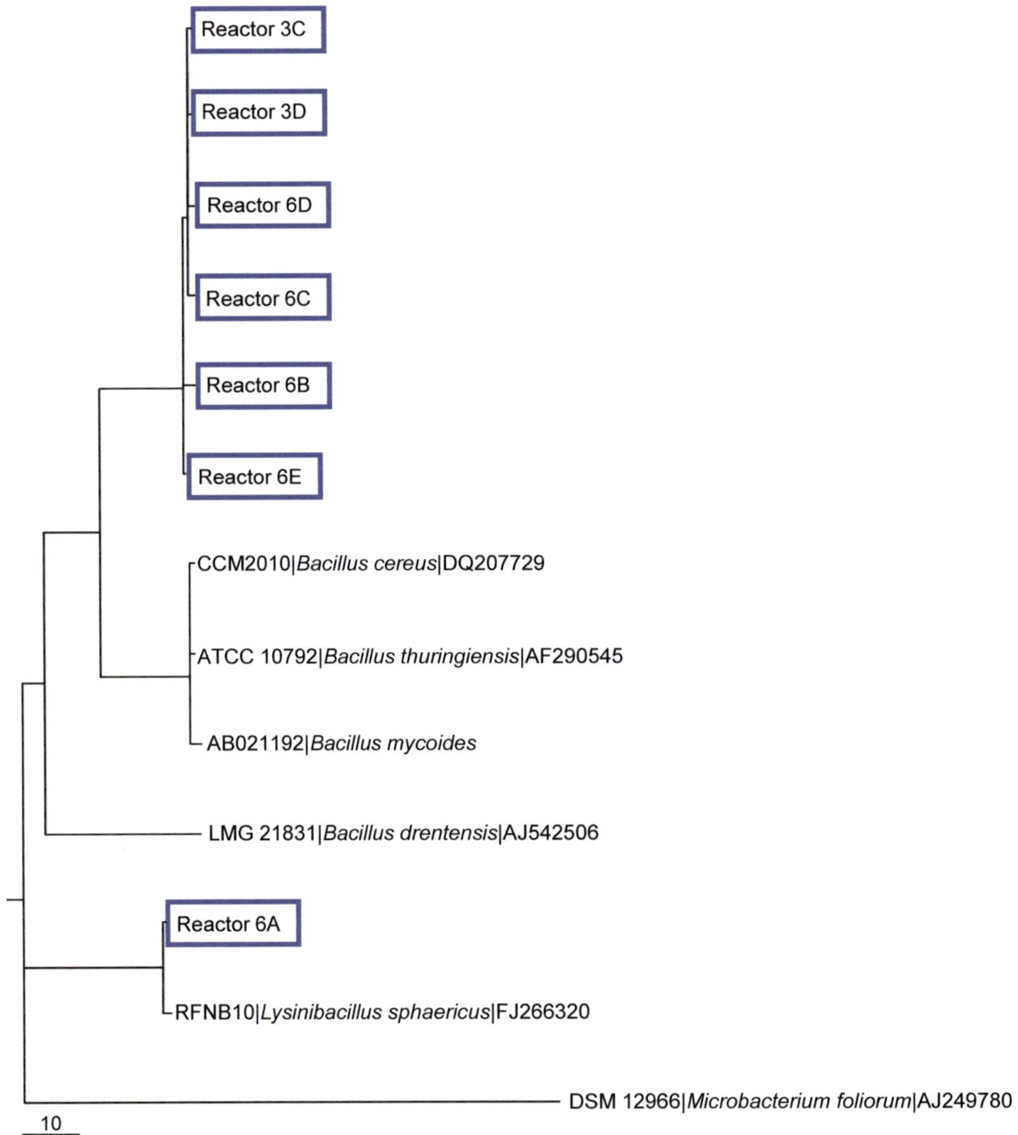

Fig. 10. Phylogenetic tree of species from microcosm reactors after operation under 40 mg/L influent Cr(VI) concentration for 17-20 days.

11. Conclusion

Since the first Cr(VI) reducing bacteria were isolated in the 1970's, a lot of progress has been made in isolating and developing higher performing cultures adapted to various environments. New research using genetic tools has yielded new cultures and new understanding of the Cr(VI) reduction process both at the molecular level [through genetic studies] and at culture community level [through genomics and proteomics]. Pure and mixed cultures of bacteria have been applied successfully in treating industrial effluents containing high levels of Cr(VI). However, application of biological systems in the remediation of contaminated environments still faces a challenge. Although culture performance under natural conditions has been evaluated using laboratory microcosms, more research is still required to elucidate the fate and possibility of recovery of artificial microbial barriers. The question of the fate of reduced Cr species and what to do about the foreseeable blockage by hydroxide species remains unanswered. In order for the *in situ* bioremediation technology to work for Cr(VI) and other toxic heavy metals, a solution must be found for feasible recovery of the barrier zones involving remobilisation of reduced Cr species.

12. Acknowledgement

The research on biological Cr(VI) reduction at the University of Pretoria was funded by the National Research Foundation (NRF) of South Africa through the Focus Areas Programme Grant No FA2007030400002 awarded to Evans M. N. Chirwa.

13. References

Ackerley, D.F.; Gonzalez, C.F.; Park, C.H.; Blake R.; Keyhan, M. & Matin A. (2004). Chromate-Reducing Properties of Soluble Flavoproteins from *Pseudomonas putida* and *Escherichia coli. Applied And Environmental Microbiology*, Vol.20, No.2, (February 2004), p. 873-882, ISSN 0099-2240.

Beszedits, S. (1988). Chromium Removal from Industrial Wastewaters, In: *Chromium in the Natural and Human Environments*, pp. 232-263, Nriagu, O. & Nieboer E. (Eds.), John Wiley, ISBN 978-0471856436, New York, New York, USA.

Bopp, L.H.; Chakrabarty, A.M. & Ehrlich, H.L. (1983). Chromate Resistance Plasmid in *Pseudomonas fluorescens. Journal of Bacteriology*, Vol.155, No.3, (September 1983), pp. 1105-1109, ISSN 0021-9193.

Bopp, L.H. & Ehrlich, H.L. (1988). Chromate Resistance and Reduction in *Pseudomonas fluorescens* Strain LB300. *Archives of Microbiology*, Vol.150, No.5, (September 1988), pp. 426-431, ISSN 0302-8933.

Brown, S.D.; Thompson, M.R.; Verberkmoes, N.C.; Chourey, K.; Shah, M.; Zhou, J.Z.; Hettich, R.L. & Thompson, D.K. (2006). Molecular Dynamics of the *Shewanella oneidensis* Response to Chromate Stress. *Molecular Cell Proteomics*, Vol.5, No.3, (March 2006), pp. 1054-1071, ISSN 535-9476.

Bruhn, D.F.; Frank, S.M.; Roberto, F.F.; Pinhero, P.J. & Johnson, S.G. (2009). Microbial Biofilm Growth on Irradiated, Spent Nuclear Fuel Cladding. *Journal of Nuclear Materials*, Vol.384, No.2, (February 2009), pp. 140-145, ISSN 0022-3115.

Camargo, F.A.O.; Bento, F.M.; Okeke, B.C. & Frankenberger, W.T. (2003). Chromate Reduction by Chromium-Resistant Bacteria Isolated from Soils Contaminated with Dichromate. *Journal of Environmental Quality*, Vol.32, No.4, (July 2003), pp. 1228-1233, ISSN 0047-2425.

Cervantes, C.; Campos-Garcia, J.; Devars, S.; Gutierrez-Corona, F.; Loza-Tavera, H.; Torres-Guzman, J.C. & Moreno-Sanchez, R. (2001). Interactions of Chromium with Microorganisms and Plants. *FEMS Microbiology Review*, Vol.25, No.3, (May 2001), pp. 335-347, ISSN 0168-6445.

Chabalala, S. & Chirwa, E.M.N. (2010). Uranium(VI) Reduction and Removal by High Performing Purified Anaerobic Cultures from Mine Soil. *Chemosphere*, Vol.78, No.1, (January 2010), pp. 52-55, ISSN 0045-6535.

Chen, Y. & Gu, G. (2005). Preliminary studies on continuous chromium(VI) biological removal from wastewater by anaerobic-aerobic activated sludge process. *Bioresource Technology*, Vol.96, No.15, (October 2005), pp. 1713-1721, ISSN 0960-8524.

Chirwa, E.M.N. & Wang, Y.-T. (1997a). Hexavalent Chromium Reduction by *Bacillus sp.* in a Packed-Bed Bioreactor. *Environmental Science and Technology*, Vol.31, No.5, (May 1997), pp. 1446-1451, ISSN 0013-936X.

Chirwa, E.M.N. & Wang, Y.-T., (1997b). Chromium(VI) Reduction by *Pseudomonas fluorescens* LB300 in Fixed-Film Bioreactor. *Journal of Environmental Engineering*, Vo.123, No.8, (August 1997), pp. 760–766, ISSN 0733-9372.

Chirwa, E.N. & Wang, Y.-T. (2000). Simultaneous Cr(VI) Reduction and Phenol Degradation in an Anaerobic Consortium of Bacteria. *Water Research*, Vol.34, No.8, (August 2000), pp. 2376-2384, ISSN 0043-1354.

Cheung, K.H. & Gu, J.D. (2007). Mechanism of Hexavalent Chromium Detoxification by Microorganisms and Bioremediation Application Potential: A Review. *International Biodeterioration and Biodegradation*, Vol.59, No.1, (January 2007), pp. 8-15, ISSN 0964-8305.

Dakiky, M.; Khamis, M.; Manassra, A. & Mer'eb, M. (2002). Selective Adsorption of Cr(VI) in Industrial Waste Water Using Low-Cost Abundantly Available Adsorbents. *Advances in Environmental Research*, Vol.6, No.4, (October 2002), pp. 533–540, ISSN 1093-0191.

Dickerson, R.E. (1980). Cytochrome c and the Evolution of Energy Metabolism. *Scientific American*, Vol.242, No.1, (January 1980), pp. 136-153, ISSN 0036-8733.

Eary, L.E. & Rai, D. (1988). Chromate Removal from Aqueous Wastes by Reduction with Ferrous Ion. *Environmental Science and Technology*, Vol.22, No.8, (August 1988), pp. 972-977, ISSN 0013-936X.

Evanko, C.R. & Dzombak, D.A. (1997). Remediation of Metals-Contaminated Soils and Groundwater. Technology Evaluation Report. Ground-Water Remediation Technologies Analysis Center, Pittsburgh, Pennsylvania, USA. Available online <http://www.clu-in.org/download/toolkit/metals.pdf>

Federal Register, (2004). Occupational Safety and Health Administration. Occupational Exposure to Hexavalent Chromium. *69 Federal Register 59404*. October 4, 2004.

Francis, C.A.; Obraztsova, A.Y. & Tebo, B.M. (2000). Dissimilatory Metal Reduction by the Facultative Anaerobe *Pantoea agglomerans* SP1. *Applied and Environmental Microbiology*, Vol.66, No.2, (February 2000), pp. 543-548, ISSN 0099-2240.

Ganguli, A. & Tripathi, A.K. (2002). Bioremediation of Toxic Chromium from Electroplating Effluent by Chromate-Reducing *Pseudomonas aeruginosa* A2 Chr in Two Bioreactors. *Applied Microbiology and Biotechnology*, Vol.58, No.3, (March 2002), pp. 416–420, ISSN 0175-7598.

Garbisu, C.; Alkorta, I.; Llama, M.J. & Serra, J.L. (1998). Aerobic Chromate Reduction by *Bacillus subtilis*. *Biodegradation*, Vol.9, No.2, (March 1998), pp. 133-141, ISSN 0923-9820.

Garrels, R.M. & Christ, C.L. (1965). In *Solutions, Minerals and Equilibria*, pp. 403-435. Harper and Row Publishers, ISBN 978-0867201482, New York, New York, USA.

Guha, H.; Jayachandran, K. & Maurrasse, F. (2001). Kinetics of Chromium (VI) Reduction by a Type Strain *Shewanella alga* under Different Growth Conditions. *Environmental Pollution*, Vol.115, No.2, (December 2001), pp. 209-218, ISSN 0269-7491.

He, M.; Li, X.; Guo, L.; Miller, S.J.; Rensing, C. & Wang, G. (2010). Characterization and Genomic Analysis of Chromate Resistant and Reducing *Bacillus cereus* Strain SJ1. *BMC Microbiology*, Vol.10:221, (2010), pp. 1-10. Available online < http://www.biomedcentral.com/1471-2180/10/221>

Hill, K.E.; Fry, J.C. & Weightman, A.J. (1994). Gene Transfer in the Aquatic Environment: Persistence and Mobilization of the Catabolic Recombinant Plasmid pDIO in the Epilithon. *Microbiology*, Vol.140, No.7, (July 1994), pp. 1555-1563, ISSN 1350-0872.

Horitsu, H.; Futo, S.; Miyazawa, Y.; Ogai, S. & Kawai, K. (1987). Enzymatic Reduction of Hexavalent Chromium by Hexavalent Tolerant *Pseudomonas ambigua* G-1, *Agricultural and Biological Chemistry*, Vo.51, No.9, (September 1987), pp. 2417-2420, ISSN 0002-1369.

Ishibashi, Y.; Cervantes, C. & Silver, S. (1990). Chromium Reduction in *Pseudomonas putida*. *Applied and Environmental Microbiology*, Vol.56, No.7, (July 1990), pp. 2268-2270, ISSN 0099-2240.

Jukes, T.H. & Cantor, C.R. (1969). Evolution of Protein Molecules, In: *Mammalian Protein Metabolism*, pp. 21-123, Munro, H.N. (Ed.), Academic Press, ISBN 9780125106047, New York, New York, USA.

Li, X. & Krumholz, L.R. (2007). Regulation of Arsenate Resistance in *Desulfovibrio desulfuricans* G20 by an *arsRBCC* Operon and an *arsC* Gene. *Journal of Bacteriology*, Vol.189, No.10, (May 2007), pp. 3705-3711, ISSN 0021-9193.

Li, X. & Krumholz, L.R. (2009). Thioredoxin Is Involved in U(VI) and Cr(VI) Reduction in *Desulfovibrio desulfuricans* G20. *Journal of Bacteriology*, Vol.191, No.15, (August 2009), pp. 4924-4933, ISSN 0021-9193.

Llovera, S.; Bonet, R.; Simon-Pujol, M. & Congregado, F. (1993). Chromate Reduction by Resting Cells of *Agrobacterium radiobacter* EPS-916. *Applied and Environmental Microbiology*, Vol.59, No.10, (October 1993), pp. 3516-3518, ISSN 0099-2240.

Lloyd, J.R. (2003). Microbial Reduction of Metals and Radionuclides. *FEMS Microbiology Reviews*, Vol.27, No.2-3, (2003), (June 2003), pp. 411-425, ISSN 0168-6445.

Merian, E. (1984). Introduction on Environmental Chemistry and Global Cycles of Arsenic, Beryllium, Cadmium, Chromium, Cobalt, Nickel, Selenium, and Their Derivatives. *Toxicological and Environmental Chemistry*, Vol.8, (1984), pp. 9-38, ISSN 0277-2248.

Mertz, W. (1974). Chromium as a Dietary Essential for Man, In: *Trace Elements Metabolism*, ed. Hoekstra, W. G., Suttie, J. W., Ganther, K. E., and Mertz, (Eds.), *Proceedings of the Second International Symposium of Trace Elements Metabolism in Animals*, pp. 185-198, University Park Press, Baltimore, Maryland, USA.

Molokwane, P.E. & Chirwa, E.M.N. (2009) Microbial Culture Dynamics and Chromium (Vi) Removal in Packed-Column Microcosm Reactors. *Water Science and Technology*, Vol.60, No.2, (July 2009), pp. 381-388, ISSN 0273-1223.

Molokwane, P.E.; Nkhalambayausi-Chirwa, E.M. & Meli, K.C. (2008). Chromium (VI) Reduction in Activated Sludge Bacteria Exposed to High Chromium Loading: Brits Culture (South Africa). *Water Research*, Vol.42, No.17, (October 2008), pp. 4538-4548, ISSN 0043-1354.

Myers, C.R.; Carstens, B.P.; Antholine, W.E. & Myers, J.M. (2000). Chromium(VI) Reductase Activity is Associated with the Cytoplasmic Membrane of Anaerobically Grown *Shewanella putrefaciens* MR-1. *Journal of Applied Microbiology*, Vol.88, No.1, (January 2000), pp. 98-106, ISSN 1364-5072.

NAS (1974). The Relation of Selected Trace Elements to Health and Disease, In: *Geochemistry and the Environment - I.*, pp. 533. US National Academy of Engineering, Washington DC, USA.

Ngwenya N. and Chirwa E.M.N. (2011). Biological Removal of Cationic Fission Products from Nuclear Wastewater. *Water Science and Technology*, Vol.63, No.1, (January 2011), pp. 124-128, ISSN 0273-1223.

Nkhalambayausi-Chirwa, E.M. & Wang, Y.-T. (2001). Simultaneous Chromium(VI) Reduction and Phenol Degradation in a Fixed-Film Coculture Bioreactor: Reactor Performance. *Water Research*, Vol.35, No.8, (August 2001), pp. 1921-1932, ISSN 0043-1354.

Nkhalambayausi-Chirwa, E.M. & Wang, Y.-T. (2005). Modeling Cr(VI) Reduction and Phenol Degradation in a Coculture Biofilm Reactor. *ASCE Journal of Environmental Engineering*, Vol.131, No.11, (November 2005), pp. 1495-1506, ISSN 0733-9372.

Ohtake, H.; Fujii, E. & Toda, K. (1987). Reduction of toxic chromate in an industrial effluent by use of a chromate-reducing strain of *Enterobacter cloacae*. *Environmental Technology Letters*, Vo.11, No.7, (July 1990), pp. 663-668, ISSN 0143-2060.

Pal, A.; Sumana Dutta, S. & Paul, A.K. (2005). Reduction of Hexavalent Chromium By Cell-Free Extract of *Bacillus sphaericus* AND 303 Isolated from Serpentine Soil. *Current Microbiology*, Vol.51, No.5, (November 2005), 327-330, ISSN 0343-8651.

Park, C.H., Keyhan, M., Wielinga, B., Fendorf, S. & Matin, A. (2000). Purification to Homogeneity and Characterization of a Novel *Pseudomonas putida* Chromate Reductase. *Applied and Environmental Microbiology*, Vol.66, No.5, (May 2000), pp. 1788-1795, ISSN 0099-2240.

Park, C. H.; Gonzalez, C. F.; Ackerley, D. F.; Keyhan, M. & Matin, A. (2002). Molecular Engineering of Soluble Bacterial Proteins with Chromate Reductase Activity, In: *Remediation and Beneficial Reuse of Contaminated Sediments*, pp. 103–111., Hinchee, R.E. Porta, A. & Pellei, M. (Eds.), Batelle Press, ISBN 978-1574771299, Columbus, Ohio, USA.

Parmar, N.; Warren, L.A.; Roden, E.E. & Ferris, F.G. (2000). Solid Phase Capture of Strontium by The Iron Reducing Bacteria *Shewanella alga* Strain BrY. *Chemical Geology*, Vol.169, No.3-4, (September 2000), pp. 281–288, ISSN 0009-2541.

Romanenko, V.I. & Koren'kov, V.N. (1977). A Pure Culture of Bacteria Utilizing Chromate and Dichromate as Hydrogen Acceptors in Growth under Anaerobic Conditions. *Mikrobiologiya*, Vol.46, pp. 414-417, ISSN 0026-3656.

Shen, H. & Wang, Y.T. (1993). Characterization of Enzymatic Reduction of Hexavalent Chromium by *Escherichia coli* ATCC 33456. *Applied and Environmental Microbiology*, Vol.59, No.11, (November 1993), pp. 3771-3777, ISSN 0099-2240.

Shen, H. & Wang Y T. (1994). Modeling Hexavalent Chromium Reduction in *Escherichia coli* ATCC 33456. *Biotechnology and Bioengineering*, Vol.43, No.4, (April 1994), pp. 293-300, ISSN 0006-3592.

Shen, H., Pritchard, P.H. & Sewell, G.W. (1996). Microbial Reduction of Cr(VI) during Anaerobic Degradation of Benzoate. *Environmental Science and Technology*, Vol.30, No.5, (April 1996), pp. 1667-1674, ISSN 0013-936X.

Stoodley, P.; DeBeer D.; Boyle, J.D. & Lappin-Scott H.M. (1999). Evolving Perspectives of Biofilm Structure. *Biofouling*, Vol.14, No.1, (January 1999), pp. 75-94, ISSN 0892-7014.

Thacker, U.; Parikh, R.; Shouche, Y. & Madamwar, D. (2006). Hexavalent Chromium Reduction by *Providencia sp.*. *Process Biochemistry*, Vol.41, No.6, (June 2006), pp. 1332-1337, ISSN 1359-5113.

Vidic, R.D. & Pohland, F.G. (1996). Treatment Walls, *Technology Evaluation Report TE-96-01*, Ground-Water Remediation Technologies Analysis Center, Pittsburgh, PA, USA. Available online <http://www.clu-in.org/download/remed/tmt_wall.pdf>

Vincze, E. & Bowra, S. (2006). Transformation of Rhizobia with Broad-Host-Range Plasmids by Using a Freeze-Thaw Method. *Applied and Environmental Microbiology*, Vol.72, No.3, (March 2006), pp. 2290–2293, ISSN 0099-2240.

Wang, P.C.; Mori, T.; Komoril, K.; Sasatsu, M.; Toda, K. & Ohtake, H. (1989). Isolation and Characterization of an Enterobacter cloacae Strain that Reduces Hexavalent Chromium under Anaerobic Conditions. *Applied and Environmental Microbiology*, Vol.55, No.7, (July 1989), pp. 1665–1669, ISSN 0099-2240.

Zakaria, Z.A.; Zakaria, Z.; Surif, S. & Ahmad, W.A. (2007). Biological Detoxification of Cr(VI) using Wood-Husk Immobilized *Acinetobacter haemolyticus*. *Journal of Hazardous Materials*, Vol.148, No.1-2, (September 2007), pp. 164-171, ISSN 0304-3894.

Zhiguo, H.; Fengling, G.; Tao, S.; Yuehua, H. & Chao, H. (2009). Isolation and Characterization of a Cr(VI)-Reduction *Ochrobactrum sp.* strain CSCr-3 from Chromium Landfill. *Journal of Hazardous Materials*, Vol.163, No.2-3, (April 2009), pp. 869-873, ISSN 0304-3894.

Zhu, W.; Chai, L.; Ma, Z.; Wang, Y.; Xiao, H. & Zhao, K. (2008). Anaerobic Reduction of Hexavalent Chromium by Bacterial Cells of *Achromobacter sp.* Strain Ch1. *Microbiological Research*, Vol.163, No.6, (November 2008), pp. 616-623, ISSN 0944-5013.

Part 3

Biodiversity Measures

6

Biodiversity Measures in Agriculture Using DNA

Lucia Vieira Hoffmann[1], Tereza Cristina de Oliveira Borba[2], Laísa Nogueira Allem[3], Paulo Augusto Vianna Barroso[1] and Raquel Neves de Mello[2]

[1]Embrapa Cotton
[2]Embrapa Rice and Beans
[3]MSc, Post graduate fellow, National Council for Scientific and Technological Development (CNPq)
Brazilian Agricultural Research Corporation, Brazil

1. Introduction

How to measure biodiversity? One of the possibilities is to use DNA. The mention of DNA can hint evolution, but this question is much more complicated and beyond the proposition of a biodiversity measure. The reasons stated for use DNA instead of other possible molecules could be that it is stable and responsible for the transmission of traits to future generations. But another reason is simple that it is suitable for measures. First, due to universality among all living things. Second, because it is a big molecule, constituted by variations of just four bases, making it easy to make a great number of sequence comparisons and used it on biodiversity measures. When individuals to be compared are similar, a greater number of comparisons have to be made to obtain a quantification of the differences among them. Conservation biology is not concerned only with the extinction of species, but also with diversity within species or subspecies, where accurate measures of diversity may be required. A third reason is that a good amount of methodologies is already developed to study its nature.

DNA markers when used to study biodiversity are frequently designed for a great number of comparisons of similarities or differences among individuals, groups of individuals, or populations. DNA is the code for protein synthesis, but in most of the studies DNA markers are considered exclusively for comparison, which may not to be linked or recognized to be linked with any trait or adaptability of the individual.

There are lots of kinds of markers, and one important thought to have in mind is that there is not a best one among them, but the choice should consider the species to be studied and the objective of the diversity analysis.

2. From nature to agriculture: why we need diversity?

The different crops and/or livestock, as we know, are indispensable components for Humankind daily routine and were domesticated in the course of modern agriculture development. The initials "attempts" of agriculture, which leaded to plants domestication, were mainly intuitive, the selection of seeds and animals were based in traits that best suited the demand from each civilization. But as soon as the initials "attempts" of agriculture were taken, guiding early "breeding" purposes, the losses of diversity became more and more apparent. After XIX century, these losses tended to increase in a higher speed, since breeder's effort in developing modern, high yield cultivars guided a substitution of wild relatives of modern crops, landraces and local varieties.

The debate about diversity, germplasm conservation and agriculture is not an original topic within breeders, germplasm bank curators, ecologists and others. Even though, each expertise embraces a different perspective about diversity. It is well known that low genetic diversity, within main crops, are due the intensive selection practiced by breeding programs, which are inclined to use few genitors (Donini et al., 2000). Such custom may leave them more vulnerable to a broad set of biotic and abiotic stresses, if not it leads to a decline in genetic variability, reducing the chance of selecting new allelic combinations (Borba et al., 2009). And since genetic variability is the raw material for selection, it is prudent to maintain it at an adequate rate to consent new combinations and therefore the exploitation of "new" desirable traits.

The competent maintenance of diversity within crops and their wild relatives, as well as plants with social-economic potential interest, is strategic not only to breeding programs but to pharmaceutical and biofuels industries, food security among others causes. But despite the immense potential that diversity holds for humankind, its unknown value comprises the major risk factor for its irreparable losses. Among the most prominent causes of diversity losses are the high rates of demographic growth and, as a result, the quick devastation of natural resources (Nass, 2001). Tropical and sub-tropical countries, which hold the greatest proportion of biodiversity (Figure 1, Figure 2), are the ones that undergo higher rates of natural devastation.

Fig. 1. Sampling of common bean (*Phaseolus vulgaris*) diversity. Archives of Embrapa Rice and Beans, Francisco Lins, 1999.

Despite the great proportion of diversity concentrated in a relative small number of countries, no country is self-sufficient in biodiversity. Therefore, the preservation of biodiversity sources, as genetic resources, is the key factor to satisfy nowadays and future needs. Different strategies are available for the maintenance of genetic resources (as a living stock of diversity), the two most "popular" are *ex situ* and *in situ* methods. Both strategies are required and important and the choice for the most suited method must reflect the species characteristics and needs.
The strategies for genetic diversity conservation can be practiced as management strategies and may vary according to the characteristics of the various plant species. Even though different species demand different conservation strategies, there are some few ordinary steps that, if followed, may guide a successful conservation of diversity. Prospection, evaluation and characterization, interchange, regeneration among others are essential for the adequate maintenance of diversity represented by genetic resources.
Along with the available tools for the management of diversity there are molecular markers, or "observing" DNA fragments which can be associated to genetic heritable traits.

Fig. 2. Sampling of cultivated rice (*Oryza sativa*) diversity. Archives of Embrapa Rice and Beans, Francisco Lins, 1999.

3. Microrganisms and pests associated to plants

Plants shoots and roots are constantly exposed to pests and microorganisms. In soil the various microorganisms frequently starve, and are nurtured and attracted by root exudates. Microorganisms may be symbionts, in an intimate association where important new morphological structures are created, at least at the cellular level; or promote plant growth by, for example, producing beneficial substances or being antagonistic to pathogens (Araújo et al., 2001; Silveira & Freitas, 2007; Torres et al., 2009). Microorganisms may be pathogens, and plants have to defend themselves. Defense can be constitutive or be triggered only by the contact with the microorganism. When defense reactions are elicited by the presence of pathogens, a system of recognition is necessary, usually performed by cell membrane proteins. Those recognition proteins are specific to pathogens species or races. Constitutive mechanisms can be much more general.

Various research efforts are directed to study important pest or pathogens diversity. It has been frequently performed with markers not related to any function or pathogenicity (Krause-Sakate et al., 2001; Ribeiro et al., 2003). The genetic structure of a pest population is probably related to geographical distances and physical barriers, and may be dependent of alternative hosts off the insect (Cunha et al., 2010). More recently, efforts have been made to identify genes related to pathogenicy and study of diversity is conducted directly with them. This seem to be the ideal measure, when the diversity study is conducted with agronomic purposes, for example, to evaluate the distribution and variability of the pathogen and its virulence effectors, as a way to infer if a given resistance gene, or some resistance genes, would be enough to control the disease. Effectors genes that encode proteins secreted in the host plants have been used to study a soil fungi diversity (Chakrabarti et al., 2011).

Plant genes responsible to resistance to pathogens have been also localized or cloned. Molecular mapping is a use of molecular markers slightly different from the study of the genetic diversity per se, because it has the aim to localize a marker physically linked in the chromosome to a gene that is responsible for a given trait. The link implies that gene and marker recombine the least when gametes are formed, causing gene and marker to cosegregate. The ideal plant population to localize a marker linked to a gene is the offspring of plants derived from a cross between paternal lines which differ specially on the trait to be mapped. Some traits are controlled by various loci, each one contributing a small amount to a quantitative trait, and are called quantitative trait loci (QTL). Plant resistance to pathogens is interesting to be mapped because there are a small number of genes, or a single gene, responsible for the trait (St Clair, 2010), which can be considered qualitative. Furthermore, the general protein structure and protein sequences of various plant resistance genes are conserved among plants, particularly those related to pathogen recognition, and new genes can be isolated by similarity (Bakker et al., 2011).

A series of markers linked to disease resistance genes are available in the literature, and they are useful in breeding programs where plants bearing the marker are selected with the aim to select resistant plants. There is special advantage of using this indirect selection, called marker assisted selection (MAS) is due to the difficulty of inoculating the high number of plants to be selected in a breeding program. Some viruses have to be inoculated through insect vectors, impracticable with a considerable amount of plants. Most of breeding programs are based on selection during natural infection, with the conduction of the experiments in conditions that favor the disease spread. But disease spread is not uniform, and plant by plant selection is sometimes required.

The adaptability of plants introduced to different environment can be improved if a selection for the resistance to particular stressed condition is performed by molecular markers. For example, the introduction of Latin American cassava genotypes to Africa has been more successful when a previous selection to the resistance to the Africa Cassava Mosaic Virus was made (Okogbenin et al., 2007).

4. Genetic markers and molecular markers

Along with the available tools for the management of diversity there are molecular markers, which may assist, reliably, the determination and examination of diversity, its conservation and, satisfactorily, guide its exploitation. Within the applications of molecular markers is the determination of how the genetic structure of a certain population, or an assemblage of germplasm accessions, is organized. The genetic structure may answer questions about how much diversity such germplasm assemblage holds, or how such genotypes must "react" under natural or artificial selection. Besides, the information resultant from molecular markers' analysis and from biometrics tools might result in the identification of novel marker alleles linked to genes involved in the expression of important traits, which can be extensively explored during cultivar development in breeding programs.

The definition of a genetic marker is not new, it was first given when the concern was not to study diversity, but to understand cosegregation of agronomic interesting traits to others characteristics of the genome, which is known by QTL and genetic mapping. Therefore, a genetic marker is defined as a heritable characteristic that can be associated to an interesting trait. When we do not think in mapping, but in diversity, the genetic marker is any genetic characteristic that is variable, or polymorphic, among the individuals to be studied, and heritable.

The genetic marker can be morphological or molecular (biochemical or DNA/RNA based). The presence or size of a spot in a flower is a morphological marker. The main advantage of molecular markers is that they can be obtained in a virtually infinite number. Furthermore it is not influenced by the environment, as parts of the morphological traits. Small organisms, as bacteria, are practically impossible to be studied through naked eye, or sometimes even with a microscope, consequently difficult to characterize morphological differences, therefore molecular markers can help. Others morphological markers can be assumed to have a relevance greater than the deserved. For example, the traditional cotton *Gossypium barbadense*, which used to be cultivated by native South American inhabitants, was classified in different subspecies when the seed from the same boll were adhered to each other, forming the called kidney seeds. This trait is, presumably, controlled by a single locus (Almeida et al., 2009), but for some authors this single trait is not relevant enough to differentiate subspecies, and molecular markers could be used to explain this process.

Population genetics has been markedly based on studies using neutral molecular markers, and the obtained genetic structure provides information individuals in a population are more related among themselves than with individuals of other populations. It is also possible to conduct population genetic studies based on QTLs or markers linked to any characteristic known to have been selected. The comparison can elucidate relative roles of selection and neutral evolution (Edelaar & Bjorklund, 2011; Stinchcombe & Hoekstra, 2008).

Monitoring forest maintenance by satellites has been criticized because it would not be enough to measure the size of the preserved forest area, but real diversity is not perceived. Species identification must be done *in situ* (Fonseca et al., 2008), and molecular markers distributed along the genome and not linked to the special selected traits may provide general diversity measures as the number of alleles per locus, as well as the population structure (Laurentin, 2009).

5. How measures are taken: a brief review on the simplest and most popular tools

For some time, sequencing was laborious and expensive, and differences among DNA molecules were accessed mainly by DNA fragment size. The amplification of DNA in vitro, or PCR, was an essential methodology to develop DNA markers. The separation can be

carried out by electrophoresis: short molecules migrate faster than long ones. Other tools are restriction enzymes, which, in nature, are enzymes synthesized by bacteria to break infecting virus DNA. Some of them make their cuts in special definite DNA sequences. The precision and reproducibility of the sequence recognition were useful on the recognition of specificities - maintenance and differences among DNA of various individuals - and so, on the development of markers.
We here briefly list some of the most used techniques to obtain markers, focusing not in the methodology but the characteristics of facility of obtaining data and ability to detect polymorphism.

5.1 Random amplified DNA reveals polymorphism

Random Amplified Polymorphic DNA (RAPD) is a friendly marker which compares individuals based on suitability for amplifications which depends on DNA complementation to random small DNA sequences. It has been used largely, but is criticized due to the low repeatability or reproducibility.
It is easy to use, and cheap, because it is based only on a PCR amplification followed by agarose gel electrophoresis. The random small DNA sequences (usually from eight to ten bases long) are used as primers of the PCR reaction. They are smaller to the oligonucleotides used in regular PCR, which are specific, therefore having a greater chance to anneal to any genome: since annealing occurs by complementarity of adenine to thiamine and of guanine to cytosine, the chance to a small sequence to find by chance a complementary sequence in a genome is relatively high. The regular PCRs are performed with longer oligonucleotides as primers therefore to the amplification is specific, chosen by the researcher, and the sequence of the oligonucleotides to be used have to be previously known.
After electrophoresis, DNA is stained, and the differences among individuals are observed as presence or absence of bands (Figure 3). Homozygous and heterozygous individuals cannot be distinguished, and the progeny of intercrossed heterozygous individuals segregates in a 3:1 proportion, therefore RAPD is a dominant marker.

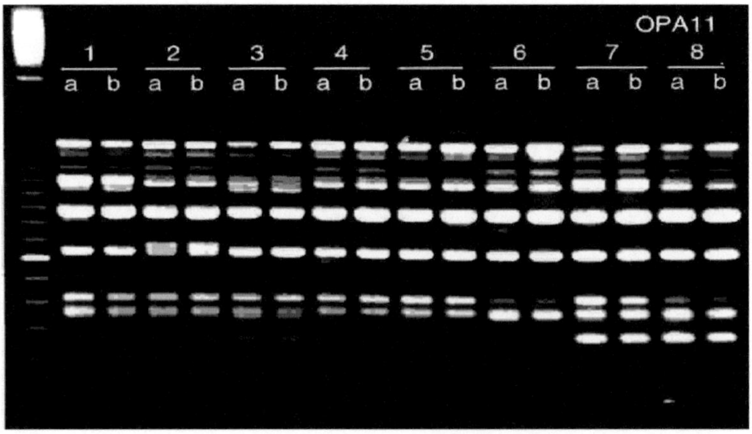

Fig. 3. Eight soybean individuals amplified with the RAPD primer OPA11. Polymorphism is noticed as presence or absence of stained bands, resulting from the number of times this primer annealed in each one of the particular genomes, with production of amplified DNA fragments separated by size in the agarose gel.

The lack of necessity of any previous knowledge of the species DNA was a great advantage of this marker.
Other techniques were developed that did not require any previous information of the species DNA, although with higher reproducibility, as AFLPs (Vos et al., 1995; Zabeau & Vos, 1993) and DART (Akbari et al., 2006; Amorim et al., 2009; Varshney et al.,2010; Wenz et al., 2004).
The RAPD markers have been independently developed by Willians et al. (1990) and Welsh e McClelland (1990).
The use of the marker in plant diversity has been reviewed by Arif et al. (2010).

5.2 Repetitive sequences can be especially polymorphic

Despite the extent of DNA molecule and the differences among individuals, it may be difficult to access variability, especially when studying genetically related individuals. In those cases, sequences with the greatest contrasts are desirable.
Microsatelites or SSR are small repetitive DNAs which are used due to be hypervariable. The repetitive bases are one to six, repeated a few times until around a hundred times, flanked by normal non repetitive sequences. The primers to reveal SSR loci are designed to be complementary to these non repetitive flanking sequences - so the disadvantage of SSR markers is the necessity of knowing the sequences before the primers design. SSR markers are more frequently in no coding regions of the genome, those which will not be transcribed or translated into DNA or proteins (Victoria et al., 2011). SSR are used multialellic, what means that for a single locus more than two forms of the marker, composed by various number of repetitions, may be found (Figure 4). Heteroziguous individuals bearing two alleles can be distinguished from any of the homozygous ones, and observing the alleles of a population derived from the crossing among two heterozygous individuals the proportion 1:2:1 can be noticed. Therefore microsatellites are codominant markers.
Frequently genomes have been sequenced not from the whole DNA of the organisms, but from expressed sequences only. For that, the initially collected material is RNA, instead of DNA. It means that the sequences obtained are all expressed, called expressed sequence tags (EST). These data can be a source to mine SSRs, and when obtained this way the SSRs belong to a expressed sequence.
Following the PCR an electrophoresis is performed to separate fragments of different sizes. Differently from RAPD, SSR of the individuals differ from others just by a few bases, so the separation has to be sharp to lead size identification. For that reason fragments are separated on acrylamide gels, not agarose gels, or capillary electrophoresis in a sequencing equipment.

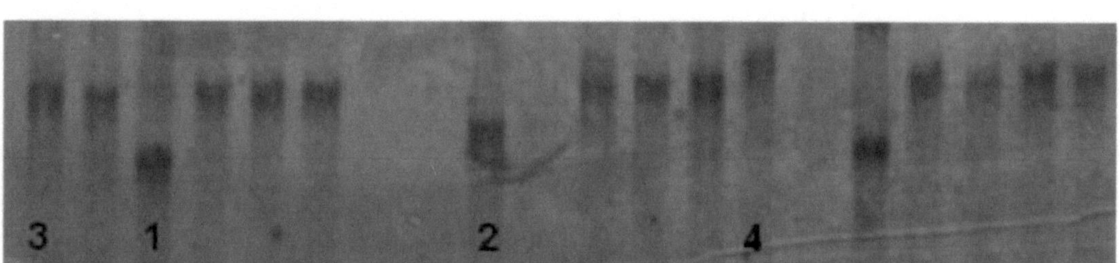

Fig. 4. Fragments of DNA resulting for the amplification with SSR primer pairs separated by size by electrophoresis in an acrylamide gel and stained by silver. For different DNA sizes are shown, corresponding to four different alleles.

5.3 Sequencing and single nucleotide polymorphism

Recently, sequencing became much more reliable, allowing the discovering of differences between individuals of a single nucleotide, or single nucleotide polymorphism (SNP). Differently from the previous cited markers, the definition of single nucleotide polymorphism is not dependent on the methodology used for their detection, which can be various. The most efficient are oligonucleotide arrays (Gupta et al., 2008).

Similarly, small insertion or deletion (INDEL) have been localized and their frequencies measured (Zhidkov et al., 2011).

Those markers are much more abundant and precise, and should turn out to be the most used, at least among the most studied species. They are the new frontier to measure the biodiversity, and have been used to study human pathogens diversity and epidemiology (Baker et al., 2010).

Independently of the marker class, working with DNA fragments may require criteria to demonstrate the reliability of the results. Sharing information by publications and websites may be very useful to verify reproducibility of results.

6. Markers may help to understand evolution

Evolution is rarely accessed experimentally, but by observation and measures taken in natural environment and inference. Hypothesis in this field may look more theoretical than in others.

It is known that natural selection depends on fitness, which may be defined as the ability to produce descendents. Fitness is dependent on the interaction with environment.

The hole of hybridization in evolution has been despised since a various interespecific hybrids present smaller general development and reduced or absent seed production. Molecular markers have lead to show that well established plants are hybrids (Ellstrand, 2003) and may have supplanted their parents (Hegde et al., 2006).

Genetic drift may have importance in evolution, which can be understood by loose of variability, caused for example by death of huge amount of population individuals due to natural phenomena or human actions. It is not unusual that a plant species suffer with an environmental or disturbance by human action causing a marked reduction of the population size. Afterwards, the remaining individuals reproduce so size of the population is recovered, but not with the ancient diversity. This phenomena is called a genetic bottleneck, and molecular markers are able to track them by identifying a population with great number of individuals with genetic diversity smaller then a small population of the same species (Barroso et al., 2010). The smallest diversity reveals disturbance among wild plants.

The importance of genetic drift has been shown experimentally in a for years experiment with *Lolium perenne* (Nestmann et al., 2011).

7. The gains in plant breeding depends of variance

Biodiversity is important not only in nature, but also on agriculture systems. The goals of plant breeding are productive plants, resistant to draught and temperature, pathogens and insects, efficient on nutrients uptake and symbiosis, etc. Novel characteristics or use of plant species can also be a challenge, like production of biofuels (Paterson et al., 2009). The way to achieve this is to find within the species to be bred plants bearing the genes conferring the desired trait or, if not available, within related species which intercrosses with it.

The number of traits that can be introduced by genes of non related plant species by plant genetic transformation is restricted mainly by the number of genes necessary for the characteristic to be expressed. Only those controlled by a small number of genes can be introduce by genetic transformation, and usually a single gene is introduced. Difficulties on knowing useful genes, which may not have been already isolated and characterized, may also exist.

Productiveness is economically believed to be major challenge to agriculture in face of the human population growth. Plant breeding has a major hole on increase agricultural production by the development of seeds - and for that the selection have to be performed among the plants that already are productive and adapted to cultivation. The continuous procedure causes loss of general biological diversity (Bai & Lindhout, 2007) and genetic diversity, which can be noticed by a loss in allele richness.

The gains achieved by plant breeding may decrease in years of selection due to the loss of genetic richness and allele segregation within the breeding population (Campbell et al., 2010). How genetic variability could be enhanced or preserved? The introduction of the crop relatives not so adapted to the cultivation system is referred as pre breeding, which are crossed to well adapted genotypes. The low productiveness of the offspring compared to the adapted parent and the years of crossing and the years of crossings and selection necessary to recover the initial production level discourages its use. Molecular markers can help hear not to maintain diversity, but otherwise to recover the adapted parent traits, with the use of recurrent selection. The marker assisted selection when used to select to the productive parental genotype may help to recover production levels in a much lesser number of years. Selecting the crop genotype is the aid molecular markers can play to foster introduction of non adapted genotypes to plant breeding.

Colored cotton fibers exist in nature, but cotton breeders have been selected for white fibers, easier to be industrial stained (Figure 5). The development of color cotton varieties avoids environmental pollution caused by staining (Teixeira et al., 2010).

Because breeding programs are expensive, and a great number of the populations which are conducted may not produce interesting seeds of varieties, models have been developed to use the evaluation by molecular data of candidate parents for prediction of the performance of the population resulting from their crossings (Barroso et al., 2003).

Fig. 5. (Continued)

Fig. 5. *Gossipium mustelinum*, a native cotton species endemic to northeast Brazil semiarid region. While the cultivated cotton (A) retains the fiber and seeds, a trait selected by plant domestication, the seeds of the wild cotton are naturally released from the boll (B) and will be dispersed through streams. Young plants survive due the protection from goat feeding by a common thorn plant *Bromelia lacinosa* (C). Adult plants can be high (D) so animals damage but not destroy them.

8. Conclusion

We are in a period of constant innovations in methodologies to access genetic diversity, in which some methodologies in use can be seen as obsolete when faced to newly developed ones. For a number of well the best studied species, genetic diversity measures data are easily obtained and available. The use of molecular data to monitor genetic diversity lead improved understanding over evolution. The increasing amount of data of crops and their relatives should foster the actual use of genetic resources in plant breeding.

9. References

Akbari, M., Wenzl, P., Caig, V., Carling, J., Xia, L., Yang, S., Uszynski, G., Mohler, V., Lehmensiek, A., Kuchel, H., Hayden, M. J., Howes, N., Sharp, P., Vaughan, P., Rathmell, B., Huttner, E. & Kilian, A. (2006). Diversity Arrays Technology (DArT) for high-throughput profiling of the hexaploid wheat genome. *Theoretical and Applied Genetics*, Vol. 113, No. 8, (October 2006), pp. 1409-1420, ISSN 0040-5752

Almeida , V. C., Hoffmann, L. V., Yokomizo, G. K. I., Costa, J. N., Giband, M. & Barroso, P. A. V. (2009). In situ and genetic characterization of *Gossypium barbadense* populations from the states of Pará and Amapá, Brazil. *Pesquisa Agropecuária Brasileira*, Vol. 44, No. 7, (July 2009), pp. 719-725, ISSN 0100-204X

Amorim, E. P., Vilarinhos, A. D., Cohen, K. O., Amorim, V. B. O., Santos-Serejo, J. A., Silva, S. O., Pestana, K. N., Santos, V. J., Paes, N. S., Monte, D. C. & Reis, R. V. (2009). Genetic diversity of carotenoid-rich bananas evalueted by Diversity Arrays Technology (DArT). *Genetics Molecular Biology*, Vol. 32, No. 1, (n. d.), pp. 96-103, ISSN 1415-4757

Araújo, W. L., Maccheroni Jr., W., Aguilar-Vildoso, Barroso, P. A. V., C. I., Saridakis, H. O. & Azevedo, J. L. (2001). Variability and interactions between endophytic bacteria and fungi isolated from leaf tissues of citrus rootstocks. *Canadian Journal of Microbiology*, Vol. 47, No. 3, (March 2001), pp. 229-236, ISSN 0008-4166

Arif, I.A., Bakir, M.A., Khan, H.A., Al Farhan, A.H., Al Homaidan, A.A., Bahkali, A.H., Al Sadoon, M. & Shobrak, M. (2010). A brief review of molecular techniques to assess plant diversity. *International Journal of Molecular Science*, Vol. 11, No. 5, (May 2010), pp. 2079-2096, ISSN 1422-0067

Bai, Y. & Lindhout, P. (2007). Domestication and Breeding of Tomatoes: What have We Gained and What Can We Gain in the Future? *Annals of Botany*, Vol. 100, No. 5, (August 2007), pp. 1085–1094, ISSN 0305-7364

Baker, S., Hanage, W.P. & Holt, K.E. (2010). Navigating the future of bacterial molecular epidemiology. *Current Opinion in Microbiology*, Vol. 13, No.5, (October 2010), pp. 640-645, ISSN: 1369-5274

Bakker, E., Borm, T., Prins, P., Van der Vossen, E., Uenk, G., Arens, M., Boer, J., Van Eck, H., Muskens, M., Vossen, J.,Van der Linden, G.,Van Ham, R., Klein-Lankhorst, R., Visser, R., Smant, G., Bakker, J. & Goverse, A. (2011). A genome-wide genetic map of NB-LRR disease resistance loci in potato. *Theoretical and Applied Genetics*, (May 2011), ISSN 0040-5752

Barroso, P. A. V., Geraldi, I. O., Vieira, M. L. C., Pulcinelli, C. E., Vencovsky, R. & Dias, C. T. S. (2003). Predicting performance of soybean populations using genetic distances estimated with RAPD markers. *Genetics and Molecular Biology*, Vol. 26, No. 3, (n. d.) pp.343-348, ISSN 1415-4757

Barroso, P. A. V., Hoffmann, L. V., Freitas, R. B., Batista, C. E. A., Alves, M. F., Silva, U. C. & Andrade, F. P. (2010). In situ conservation and genetic diversity of three populations of *Gossypium mustelinum* Miers ex Watt. *Genetic Resources and Crop Evolution*, Vol. 57, No. 3, (August 2009), pp. 343-349, ISSN 0925-9864

Borba, T. C. O., Mendes, C. A., Guimarães, E. P., Brunes, T. O., Fonseca, J. R., Brondani, R. V. & Brondani, C. (2009). Genetic variability of Brazilian rice landraces determined by SSR markers. *Pesquisa Agropecuária Brasileira*, Vol. 44, No. 7, (July 2009), pp. 706-712, ISSN 0100-204X

Campbell, B. T., Saha, S., Percy, R., Frelichowski, J., Jenkins, J. N., Parker, W., Mayee, C. D., Gotmare, V., Dessauw, D., Giband, M., Du, X., Jia, Y., Constable, G., Dillon, S., Abdurakhmonov, I. Y., Abdukarimov, A., Rizaeva, S. M., Adullaev, A., Barroso, P. A. V., Padua, J. G., Hoffmann, L. V. & Podolnaya, L. (2010). Status of the global cotton germplasm resources. *Crop science*, Vol. 50, No.4, (July 2010), pp. 1161-1179, ISSN 0011-183X

Chakrabarti, A., Rep, M., Wang, B., Ashton, A., Dodds, P. & Ellis, J. (2011). Variation in potential effector genes distinguishing Australian and non-Australian isolates of the cotton wilt pathogen *Fusarium oxysporum* f.sp. vasinfectum. *Plant Pathology*, Vol. 60, No. 2, (September 2010), pp. 232–243, ISSN 0032-0862

Cunha, F., Gómez, D.R.S., Silva, J.J., Alexandre, T.M. & Moscardi, F. (2010). Genetic diversity of the sunflower caterpillar (*Chlosyne lacinia saundersii* Doubleday and Hewitson) (Lepidoptera: Nymphalidae) populations determined by molecular RAPD markers.

Anais da Academia Brasileira de Ciências, Vol. 82, No. 4, (n.d.), pp. 1127-1136, ISSN 0001-3765

Donini, P., Law, J.R., Koebner, R.M.D., Reeves, J.C. & Cooke, R.J. (2000). Temporal trends in the diversity of UK wheat. *Theoretical and Applied Genetics,* Vol.100, No.6, (n.d.), pp.912-917, ISSN 0040-5752

Edelaar, P. & Bjorklund, M. (2011). If FST does not measure neutral genetic differentiation,then comparing it with QST is misleading. Or is it? *Molecular Ecology,* Vol. 20, No. 9, (March 2011), pp.1805-1812, ISSN: 0962-1083

Ellstrand, N.C. (2003). Current knowledge of gene flow in plants: implications for transgene flow. *Philosophical Transactions of the Royal Society,* Vol. 358, No. 1434 , (May 2003), pp. 1163–1170, ISSN 1471-2970

Fonseca, R.M., Lopes, R., Barros, W.S., Lopes, M. T . G.& Ferreira, F. M. (2008). Morphologic characterization and genetic diversity of *Capsicum chinense* Jacq. accessions along the upper Rio Negro - Amazonas. *Crop Breeding and Applied Biotechnology,* Vol.8, No.3, (n.d.), pp. 187-194, ISSN 1518-7853

Gupta, P.K., Rustgi, S. & Mir, R.R. (2008). Array-based high-throughput DNA markers for crop improvement. *Heredity,* Vol.101, No. 1, (May 2008), pp.5-18, ISSN 0018-067X

Hegde, S. G., Nason, J. D., Clegg, J. M. & Ellstrand, N. C. (2006). The evolution of California'a wild radish has resulted in the extinc of its progenitors. *Evolution,* Vol. 60, No. 6, (n. d.), pp. 1187–1197, ISSN 0014-3820

Krause-Sakate, R., Mello, R. N., Pavan, M. A., Zambolim, E. M., Carvalho, M. G., Le Gall, O. & Zerbini, F. M. (2001). Molecular characterization of two Brazilian isolates of *Lettuce mosaic virus* with distinct biological properties. *Fitopatologia Brasileira,* Vol.26, No. 2, (n.d.), pp.153-157, ISSN 0100-4158

Laurentin, H. (2009). Data analysis for molecular characterization of plant genetic resources. *Genetic Resources and Crop Evolution,* Vol. 56, No. 2, (January 2009), pp. 277–292, ISSN 0925-9864

Nass, L.L. (2001). Utilização de recursos genéticos vegetais no melhoramento, In: *Recursos genéticos e Melhoramento – Plantas,* Nass, L.L., Valois, A.C.C., Melo, I.S. & Valadares-Inglis, M. C, (Ed.), pp.30-55, Fundação MT, Rondonópolis, Brazil

Nestmann, S., Rajicic, T. S., Dehmer, K.J., Fischer, M., Schumacher, J. & Roscher, C. (2011). Plant species diversity and composition of experimental grasslands affect genetic differentiation of *Lolium perenne* populations. *Molecular Ecology,* Vol.20, No. 10, (n.d.), pp. 2188–2203, ISSN: 0962-1083

Okogbenin, E., Porto, M.C.M., Egesi, C., Mba, C., Espinosa, E., Santos, L.G., Ospina, C., Marín, J., Barrera, E., Gutiérrez, J., Ekanayake, I., Iglesias, C. & Fregene, M.A. (2007). Marker-Assisted Introgression of Resistance to Cassava Mosaic Disease into Latin American Germplasm for the Genetic Improvement of Cassava in África. *Crop Science,* Vol.47, No. 5, (n.d.), pp.1895-1904, ISSN 0011-183X

Paterson, A.H., Bowers, J.E., Bruggmann, R., Dubchak, I., Grimwood, J., Gundlach, H., Haberer, G., Hellsten, U., Mitros, T., Poliakov, A., Schmutz, J., Spannagl, M., Tang, H., Wang, X., Wicker, T., Bharti, A.K., Chapman, J., Feltus, F.A., Gowik, U., Grigoriev, I.V., Lyons, E., Maher, C.A., Martis, M., Narechania, A., Otillar, R.P., Penning, B.W., Salamov, A.A., Wang, Y., Zhang, L., Carpita, N.C., Freeling, M.,

Gingle, A.R., Hash, C.T., Keller, B., Klein, P., Kresovich, S., McCann, M.C., Ming, R., Peterson, D.G., Mehboob-ur-Rahman, Ware, D., Westhoff, P., Mayer, K.F.X., Messing, J. & Rokhsar, D.S. (2009). The *Sorghum bicolor* genome and the diversification of grasses. *Nature*, Vol.457, No. 7229, (January 2009), pp. 551-556, ISSN 0028-0836

Ribeiro, S. G., Ambrozevícius, L. P., Ávila, A. C., Bezerra, I. C., Calegario, R. F., Fernandes, J. J. , Lima, M. F., de Mello, R. N., Rocha, H. & Zerbini, F. M. (2003). Distribution and genetic diversity of tomato-infecting begomoviruses in Brazil. *Archives of Virology*, Vol.148, No. 2, (n.d.), pp.281-295, ISSN 0304-8608

Silveira, A. P. D & Freitas, S.S. (Eds.). (2007). *Microbiota do Solo e Qualidade Ambiental*, Instituto Agronômico Campinas, ISBN 978-85-85564-14-8, São Paulo, Brazil

St.Clair, D.A. (2010). Quantitative disease resistance and quantitative resistance Loci in breeding. *Annual Review of Phytopathology*, Vol. 48, (May 2010), pp. 247-268, ISSN 0066-4286

Stinchcombe, J.R. & Hoekstra, H.E. (2008). Combining population genomics and quantitative genetics: finding the genes underlying ecologically important traits. *Heredity*, Vol.100, No. 2, (February 2007), pp. 158-170, ISSN 0018-067X

Teixeira, E. M., Corrêa, A. N., Manzoli, A., Leite, F. L., Oliveira, C. R. & Mattoso, L. R. C. (2010). Cellulose nanofibers from white and naturally colored cotton fibers. *Cellulose*, Vol. 17, No. 3, (February 2010), pp. 595-606, ISSN 0969-0239

Torres, A. R., Cursino, L., Muro-Abad, J. I., Gomes, E. A., Araújo, E. F., Hungria, M. & Cassini, S. T. A. (2009). Genetic diversity of indigenous common bean (*Phaseolus vulgaris* L.) rhizobia from the state of Minas Gerais, Brazil. *Brazilian Journal of Microbiology*, Vol.40, No. 4, (n.d.), pp- 852-856, ISSN 1517-8382

Varshney, R. K., Glaszmann, J. C., Leung, H. & Ribaut, J. M. (2010). More genomic resources for less-studied crops. *Trends in Biotechnology*, Vol. 28, No. 9, (n.d.), pp. 452-460, ISSN 0167-7799

Victoria, F. C., Maia, L. C. & Oliveira, A. C. (2011). In silico comparative analysis of SSR markers in plants. *BMC Plant Biology*, Vol 11, No.1, (n.d.), pp. 11-15, ISSN 1471-2229

Vos, P., Hogers, R., Bleeker, M., Reijans, M., Van de Lee, T., Hornes, M., Friters, A., Pot, J., Paleman, J., Kuiper, M. & Zabeau, M. (1995). AFLP: a new technique for DNA fingerprinting. *Nucleic Acids Research*, Vol. 23, No. 21, (November 1995), pp. 4407-4414, ISSN 0305-1048

Welsh, J. & McClelland, M. (1990). Fingerprinting genomes using PCR with arbitrary primers. *Nucleic Acids Research*, Vol.18, No.24, (n.d.), pp.7213-7218, ISSN 0305-1048

Wenzl, P., Carling, J., Kudrna, D., Jaccoud, D., Huttner, E., Kleinhofs, A. & Kilian, A. (2004). Diversity Arrays Technology (DArT) for whole genome-profiling of barley. *Proceedings of the National Academy of Sciences of the United States of America*, Vol. 26, No. 101, (June 2010), pp. 9915-9920, ISSN 0027-8424

Williams, J.G.K., Kubelik, A.R., Livak, K.J., Rafalski, J.A. & Tingey, S.V. (1990). DNA polymorphisms amplified by arbitrary primers are useful as genetic markers. *Nucleic Acids Research*, Vol.18, No.22, (n.d.), pp.6531-6535, ISSN 0305-1048

Zabeau, M & Vos, P. (1993). Selective restriction fragment amplification: a general method for DNA fingerprinting. *European Patent Office*, publication 0 534 858 A1, bulletin 93/13.

Zhidkov, I., Cohen, R., Geifman, N., Mishmar, D. & Rubin, E. (2011). CHILD: a new tool for detecting low-abundance insertions and deletions in standard sequence traces. *Nucleic Acids Research*, Vol 39, No.7, (January 2011), pp.1-8. ISSN 0305-1048

7

Molecular Techniques to Estimate Biodiversity with Case Studies from the Marine Phytoplankton

Linda K. Medlin[1,2] and Kerstin Töbe[3]
[1]*UPMC Univ Paris 06, UMR 7621, LOMIC, Observatoire Océanologique, Banyuls/mer*
[2]*CNRS, UMR 7621, LOMIC, Observatoire Océanologique, Banyuls/mer*
[3]*Alfred Wegener Institute for Polar and Marine Research, Bremerhaven*
[1,2]*France*
[3]*Germany*

1. Introduction

Approximately less than 10% of the known biodiversity in the marine protistan community is known, but among the pico-fraction even less is known with new groups being discovered regularly (Kim et al. 2011). This feature of hidden biodiversity was first recognized in the bacterial community but this phenomenon is now being extended into the eukaryotic fraction. Many cosmopolitan species, which we think we can easily recognize, are now being shown to be species complexes with little or no morphological markers to separate them. Spatial and temporal variation in their abundance and distribution in these complexes are also unknown. With new molecular and analytical techniques, our knowledge of marine species level biodiversity begins to unfold to understand how marine biodiversity supports ecosystem structure, dynamics and resilience. With these techniques, we can augment our understanding of biodiversity and ecosystem dynamics in all areas of the planktonic community, not just the photosynthetic ones. We will review selected molecular techniques and provide case studies to illustrate the use of these techniques.

For the past 30 years scientists have recognised that understanding and preserving biodiversity is one of the most important global challenges facing the world today. There is a science plan for Europe to address the problems associated with a potential loss of biodiversity in the marine environment, which was formulated in 1999 by the Association of Marine Science Institutes.

Biodiversity is strongly affected by the rapid and accelerating changes in the global climate, which largely stem from human activity. There is now common agreement that the world must generate plans to conserve and protect biodiversity to prevent rampant savaging for natural resources. How biodiversity is perceived and maintained affects ecosystem functioning and how the goods and services that ecosystems provide to humans can be used. Recognizing biodiversity at all levels is essential to preserving it. Terrestrial and marine ecosystems are inherently different and the management of their biodiversity requires very different approaches. Often terrestrial ecosystem generalizations concerning biodiversity patterns on both global and regional scales, the processes determining these patterns (Gosling 1994), and the resulting biodiversity loss are extrapolated to marine ecosystems. However, these extrapolations are generally incorrect because the marine environment experiences many more disturbances than their terrestrial counterparts and their dispersal patterns are not the same (Killian & Gaines, 2003). Medlin and Kooistra (2010) summarized the following fundamental differences between marine and terrestrial biodiversity. The physical environment in the oceans is three dimensional, whereas on land it is essentially two-dimensional. The vast majority of the biomass of marine primary producers is composed of minute and usually mobile micro-organisms, with representatives from most of the eukaryotic crown lineages (sessile macroalgae are only minor players), whereas on land, the bulk of the primary production is carried out by macroscopic and sessile green plants. Climax communities never develop in the ocean as they were once believed to have developed on land. In the ocean, primary production is consumed daily, but on land, most primary production enters the detrital cycle each autumn. Higher-level carnivores often play key roles in structuring marine biodiversity and when exploited heavily, as in over-fishing, there are severe downward-cascading effects on biodiversity and on ecosystem functions. Marine systems are more open than terrestrial and dispersal of species occurs over much larger ranges than on land (Killian & Gaines, 2003). Life originated in the sea and thus has a much longer evolutionary history in the sea than on land (Ormond et al., 1998). There are 14 indigenous marine animal phyla, whereas only one phylum is unique to land, making diversity at higher taxonomic levels higher in the sea. Four new algal phyla have been described in the last twenty years (Moestrup, 1991, Andersen et al., 1993, Guillou et al., 1999, Kawachi et al., 2002). Three new pico-sized classes await formal descriptions (Tomas et al., unpublished, Not et al., 2007, Kim et al., 2011). The sum total of genetic resources in the sea is therefore inferred to be much more diverse in the sea than on land (Grassle et al., 1991). Also on average, genetic diversity within a species (i.e. below the species level) is higher in marine than in terrestrial species. Thus, because of these fundamental differences, our understanding of marine biodiversity lags far behind that of terrestrial biodiversity. There is not enough scientific information to design management and conservation plans for the sustainable use of coastal resources.

Biodiversity can be described in three hierarchical levels: genetic, species, ecosystems. Each has its own spatial scale from single samples to regional and global populations, and temporal scales changing from short time intervals (days to weeks) to long (years to decades). On land, the full range of these scales can be sampled, but not in the ocean. In the ocean the planktonic population that is sampled at any one point in time will not be same population at that location the next day. Each scale can be affected by loss but loss at any of these scales is rarely calculated and the knock-on effect of any loss at one scale to another scale is unknown. Marine biodiversity is more widely commercialized than that on land because of the many species used as food stocks, whereas fewer species are used as food stock in terrestrial ecosystems. Exploitation of marine biodiversity is not well regulated and harvesting and fishing technology is so advanced that many marine species are now driven to extinction or near extinction.

Global biodiversity projects must first characterize the existing biodiversity as fully as possible (from genetic to ecosystem level) in selected key (flagstone) habitats across broad geographical ranges. However, this is a monumental task to compile comprehensive inventories even at a few sites. The Census of Marine Life (*Http://www.coml.org/*) is a global network of researchers from over 70 countries that tries to answer the questions "What lived in the oceans?" "What lives in the oceans?" and "What will live in the oceans?" Molecular methods have proven to an indispensable tool to answer these questions.

The world's oceans cover 70 percent of the Earth's surface, and their dominant populations, both numerically and biomass-wise, belong to microscopic protists and prokaryotes. The marine phytoplankton are major components of both these groups and are assumed to be high dispersal taxa with large population sizes. Small photosynthetic organisms are responsible for the bulk of primary production in oceanic and neritic waters. These organisms play pivotal roles in many biogeochemical processes that regulate our global climate. Net samples and bulk process measurements, such as chlorophyll *a* and ^{14}C biomass estimates have historically provided most of our knowledge about marine phytoplankton. However, whole water samplers and new analytical methods, e.g., flow cytometry, epifluorescence microscopy and HPLC (high pressure liquid chromatography) have found previously unrecognised groups (such as *Prochlorococcus*), size classes (the picoplankton < 3 µm) and hidden biodiversity (new algal classes, e.g., Bolidophyceae, Pelagophyceae, picobiliphytes). Although the global importance of picoplankton was unknown 30 years ago, they can contribute up to 90% primary production in oligotropic oceanic waters (Waterbury et al., 1979, 1986, Chisholm et al., 1988).

Because of these recent discoveries about phytoplankton biodiversity, we must ask the questions: Do we know all of the groups in the phytoplankton? Do we know how they are related to one another? Do we know their spatial and temporal changes in their abundances? Do we know the extent of their genetic diversity? The answer to these questions is an unequivocal NO.

In picoeukaryotes, where there are far too few morphological markers explored upon which to determine species identification, α-level taxonomy is lacking. A new group of picoplankton was only discovered this year (Kim et al., 2011). In addition, we know the population structure of the phytoplankton in only a few isolated cases and many of these belonging to the toxic dinoflagellate genera. It is likely to be very different from that on land because marine planktonic organisms live in an ever-changing three-dimensional environment. Many taxa may have little genetic structure over very large geographic areas. However, where population structure has been studied in the marine phytoplankton, global populations have appeared fragmented with some adjacent areas with limited gene flow between them (see review in Medlin et al., 2000). Admittedly, most of these studies have not sampled the phytoplankton species over their entire range, but if their population are fragmented on a local scale, then by inference, they are fragmented on a global scale. Further, recent evidence suggests that speciation and dispersal mechanisms in marine planktonic organisms may be very different from those on land (Killian & Gaines, 2003).

The advent of molecular biological techniques has greatly enhanced our ability to analyse all populations (Parker et al. 1998), not just the marine phytoplankton. The small size and paucity of morphological markers of many phytoplankton species, the inability to bring many into culture, and the difficulty of obtaining samples for long term seasonal studies in open ocean environments has hampered our knowledge of phytoplankton diversity and population structure. The idea of a single globally distributed species or of temporal stasis is no longer valid. Temporal genetic change may often be greater than spatial change or change between species (Brand, 1982, 1989, Gallagher, 1980, Hedgecock, 1994) and may very well apply to bloom populations. Because the rate of genetic change can and does occur on ecological time scales (Palumbi, 1992), this suggests that mechanisms are in place to determine how local adaptations and speciation can occur in apparently homogeneous populations (Gosling 1991). Now molecular techniques can present a quantitative framework through which the diversity, structure and evolution of marine phytoplankton populations can be analyzed, predictive models of the dynamics of ocean ecosystems formulated, and the idea of functional groups in the plankton proven.

2. Determining biodiversity in environmental samples by sequence analysis

The most exact method to assess biodiversity down to the species level in environmental samples is by sequencing clones from such samples. The SSU rRNA gene is often the gene of choice for cloning and is the gene most commonly used as a phylogenetic yardstick. This is best achieved by isolating total DNA from the sample followed by full-length SSU gene amplification using PCR and universal primers, then cloning and sequencing. The method allows the exhaustive description of biodiversity in a sample down to the species level. Also the resulting sequence information may serve as a basis for developing specific oligonucleotide probes necessary for subsequent methods like FISH. It should be noted though that even universal PCR primers might only amplify a subset of all organisms and therefore bias the result. It has been shown that different groups of organisms were detected when different primers have been used and if possible the analysis of an environmental sample should always include the use of different primers to get a more complete picture of its diversity.

2.1 Clone libraries

The first assessments of ecosystem biodiversity were made using clone libraries from DNA and in every case far more diversity was revealed than expected (see review in Bull 2004). However, these early clone libraries were limited by sequencing capacity and most statistical analysis revealed that coverage of the diversity of the clones had not reached a plateau. This problem has more or less been eliminated with new age sequencing. Also clone libraries made from RNA and not DNA are not identical (Lami et al., 2009).

2.2 454 sequencing and the rare biosphere

The culture independent 454 pyrosequencing is rapidly gaining favor for environmental analysis because it allows a rapid attainment of around 400 bp in a 10-hour run from an exhaustive search of a library. This exhaustive search has revealed many sequences (operational taxonomic units, OTUs) that are represented by only a single clone in the library. With traditional methods of making and sequencing clone libraries, these single sequences would not have been recovered to a large extent. This plethora of single occurring OTUs has been termed the "rare biosphere" [Sogin et al., 2006] and much effort is now being concentrated to recover this aspect of many communities with 454 sequencing or pyrosequencing as it is often referred to. The reason for this rare biosphere is unknown but it is clear that the same species are not repeated in different geographic areas (Brazelton et al. 2010). Also this technique has enabled more genes to be explored and community analysis is now moving into the age of metagenomic and metatranscriptomic analysis (Cuvelier et al., 2010). However, until the length of the sequence read is increased, full phylogenetic assignment is not attainable.

2.3 Barcoding

The barcode is defined as a short gene sequence from a standardized region of the genome (the "barcode") that can characterize, and distinguish species, and to assign unidentified individuals to species. Basically, this method is not different from the sequencing methods mentioned before but what is new here is the scale at which international consortia and scientists try to analyze biodiversity in a standardized way. The Consortium for the Barcode of Life (*Http://barcoding.si.edu/index_detail.htm*), for example, has started initiatives to develop DNA barcodes for all fish and bird species on Earth, and many other groups of organisms are targeted the same way. The primary opposition to barcoding is that it could lead to the elimination of taxonomy but this is not justified.

For barcoding to work, the "barcoding community" must agree on the gene fragment to use so that barcodes from different species are comparable. The mitochondrial COI gene (cytochrome oxidase I) is most often used for DNA Barcoding, especially in animals, but it cannot be used in many groups of phytoplankton because of the non-specificity of primers. Therefore other gene fragments, i.e., RUBISCO and ITS have been explored (Evans et al., 2007). It is likely that DNA Barcodes will be developed using many genes. DNA Barcoding will be a powerful taxonomic tool to analyze marine biodiversity. The high-throughput sequencing approach and the comparability of data will address many questions regarding cryptic and invasive species, and to identify quickly microbial diversity in any sample. Again, the main limiting factor is that barcodes of all possible organisms in the biosphere must be determined first. Is the barcode of a single individual representative of the species because different individuals in a population, let alone individuals in different geographic populations could possess slightly different barcode sequences? So again, we need to know the extent of intraspecific variation, and this variation should remain far less than differences among species. Yet, if all these problems can be solved at least in part, barcoding provides a very powerful tool of obtaining semi-quantitative data on the species composition of e.g., large numbers of environmental samples in a rapid and cost-effective way.

3. Fingerprinting methods as applied to environmental samples

Often it is not possible or necessary to get a full assessment of biodiversity but instead it may be enough to identify temporal changes or spatial differences among samples. In this case, DNA fingerprinting methods can be used. These are several PCR based methods of determining population structure. All these methods exploit differences in the length and base composition of specific gene segments which result in different banding patterns after electrophoresis - the "fingerprint" of the sample. Many of the methods work with any sequence that can form a secondary structure. Fragments of identical size but different base composition can then be separated in either denaturing or non-denaturing polyacrylamide gels, depending on the method.

DNA polymorphisms between individuals can, e.g., be found by Restriction Fragment Length Polymorphism (RFLP), a technique in which DNA is digested by restriction enzymes and then the presence or absence of restriction sites in different individuals is compared as well as insertions or deletions in their genome between these restriction sites. A slightly different RFLP method consists of the PCR amplification of a specific gene, e.g., the SSU rDNA, followed by restriction digestion with enzyme and gel electrophoresis. Because it uses only a limited number of fragments this method avoids the need of blotting and probing for visualisation and is much faster and easier than the "classical" RFLP. On the other hand, the limited number of possible bands leads also to a very small number of possible polymorphisms and one needs luck to find a usable marker. Nevertheless, there are examples where this kind of RFLP marker has been used with success, e.g., for discriminating species and strains of the toxic dinoflagellate genus *Alexandrium* (Scholin et al., 1994a).

Two well-established methods for assessing diversity in environmental samples are Temperature Gradient Gel Electrophoresis (TGGE) and Denaturing Gradient Gel Electrophoresis (DGGE) (Muyzer et al., 1993). These methods allow the qualitative and semi-quantitative determination of biodiversity in environmental samples through amplification of a short segment of DNA that is electrophoresed either through a temperature or a density gradient gel. With either system bands of different base composition do not migrate to the same location, thus generating the sample fingerprint. Bands can be cut out of the gel and sequenced.

Single-stranded-conformation polymorphism (SSCP) uses the fact that single stranded DNA fragments fold into secondary structures depending on their base composition. Small fragments of ca. 300 bp are most useful. This method does not require gradient gels or temperature gradient electrophoresis and can be used in normal sequencers with fragment length analysis programs. Also the fact that one of the two strands is degraded reduces the variability obtained from communities because it avoids heteroduplex formation, a problem known in community analyses based on DGGE.

The first widely used PCR marker technique was Random Amplified Polymorphic DNA (RAPD) or Arbitrary Primed PCR (AP-PCR) with the former being the most commonly used name for this kind of method (Hadrys & Balick 1992, Lynch & Milligan 1994). These methods use a single short random primer in a PCR reaction, most often a decamer, to amplify the DNA, which produces a fingerprint of multiple bands and polymorphisms between individual samples are derived from single nucleotide changes that prevent or allow primer binding and therefore lead to different banding patterns between individuals. This method became quite popular because it could be carried out in a short time without previous knowledge of the organism under investigation. Nevertheless, RAPDs have been shown to have some drawbacks: The use of short primers gives not only the possibility of random binding in all kind of genomes and therefore makes this method working at all, but it also makes it unreliable, too, and susceptible even to small changes in the PCR conditions. Unfortunately, RAPD markers are hard to reproduce even within the same laboratory. Also, RAPDs are normally dominant markers by which they give less information than other, mostly co-dominant markers. RAPDs should only be used when time and resources are limited and no previous information about the species under investigation are known, otherwise other markers should be targeted. Populations of the dinoflagellate *Gyrodinium catenatum* among Australian and global populations displayed spatial and temporal differences using RAPD fingerprinting data (Boalch et al., 1999). Despite this, it was not possible to define the route of introduction into Australian waters, although the introduction is judged to be quite recent based on fossil sediment records. RAPDS were used to assess populations of *Emiliania huxleyi* (Barker et al., 1995) and was one of the first studies to show that blooms were not clonal and high diversity could be shown in relatively small bodies of water, viz., mesocosms.

Recently AFLP (Amplified Fragment Length Polymorphism) has become a popular marker technique for studying biodiversity in the marine environment. It combines the advantages of RAPDs and RFLPs into a powerful tool (Vos et al., 1995). First, genomic DNA is digested with two different restriction enzymes, a rare and a frequent cutter. Then matching adapters are ligated to the digested fragments. Afterwards, a PCR is performed with primers homologous to the adapters plus up to four additional random bases at its 3' end. By using these selective bases, only a subset of digested DNA fragments is amplified, giving distinct bands instead of a smear and making it possible to analyse the bands on a polyacrylamide gel. The major advantage of this technique is the large number of bands it produces, giving a very good chance of finding a large number of polymorphic bands among them. The polymorphisms detected by this method come from the same sources as in RFLPs, insertions, deletions and point mutations leading to the presence or absence of restriction sites, but compared to RFLPs, AFLPs are normally scored only as dominant markers, even when some researchers give possible methods for using them co-dominantly. The use of longer PCR primers that anneal to the adapters and a few bases of the genomic DNA make the whole reaction much more reliable than RAPDs, because higher annealing temperatures can be used. The greatest advantage of the RAPD technology on the other hand remains, because no previous sequence information of the species under investigation is needed and PCR reactions are fast to be carried out. Nevertheless, AFLPs are technically demanding, sensitive to the purity and quantity of DNA to be digested, need some experience to be performed and data analysis of the hundreds of amplified bands should be done by computer analysis. Since 1995 when AFLPs were first introduced, there has been an increasing number of publications using this technique, but most of them deal with population studies or the development of genetic linkage maps for higher plants. Among algae, the multicellular red alga *Chondrus crispus* was the first organism to be analyzed by AFLPs, and more seaweeds have been investigated since then (e.g., *Caulerpa, Chara* and *Porphyra* species), but the method has since then also been used for phytoplankton, e.g. the marine dinoflagellate *Alexandrium tamarense,* the diatom *Asterionella formosa* and the chlorophyte *Chlorella vulgaris,* both freshwater algae. AFLP banding patterns in isolates of the dinoflagellate *Alexandrium tamarense* from the Orkney Islands were correlated with toxin patterns as determined by HPLC analysis (John et al., 2004), but a later study in the same area with more isolates and depending on the spatial scale investigated, AFLP patterns did not correlate with allelopathic capabilities (Alpermann 2009). A preliminary study of *Phaeocystis antarctica* indicated that the gyres around the Antarctic were not isolated from one another and it was likely that the ACC provided the vehicle for dispersal around the continent (Gaebler et al., 2007).

4. Analysis of population structure using molecular markers

The first molecular markers to be used in all fields were isozymes. These are proteins that show only small differences in their size or iso-electric point and therefore can be separated by starch gel electrophoresis but are still able to catalyse the same biochemical reaction. Their advantages of quick and easy isolation and detection made them the markers of choice for many investigations. But the requirement that isozymes must still be functional in the biochemical pathways strongly limits the number of possible mutations and therefore the number of alleles and the heterozygosity of this marker type. Another disadvantage of this kind of marker is also that protein content of cells and following the detectability of isozymes is strongly influenced by the environment and as a consequence, marker types were developed that directly used environment-independent DNA.

The goal of most early molecular studies concerning microalgae using isozyme analysis was to resolve species-level issues among species with conflicting or little morphological resolution rather than to study genetic structure within bloom populations. The recognition of cryptic species or the recognition of previously discounted morphological markers that can be used for separation of a species complex was the most common results of early isozyme studies. For example, different in isozyme banding patterns in neretic, shelf and oceanic populations of *Thalassiosira pseudonana* prompted Murphy & Guillard (1976) and Brand et al. (1981) initially to suggest that this species was composed of clinal populations but later detailed morphological investigations separated each ecological population into a different species (Hasle, 1978, 1983) for *Thalassiosira guillardii, oceanica* and *pseudonana*). There are many examples among the dinoflagellates where significant insights into species complexes have been made with isozymes that show in some areas populations are unique and in others they are not (*Alexandrium tamarense/fundyense/catenella*, Cembella & Taylor, 1986, Cembella et al., 1989; Hayhome et al., 1989; *Gambierdiscus toxicus* Chinain et al., 1997 and *Peridinium volzii*, Hayhome et al., 1987). In most of these cases, the isozyme conclusions were supported by further studies with sequence analyses. *Alexandrium* species have been studied in more detail using sequence analysis of rapidly evolving genomic regions, such as the ITS and the D1/D2 region of the LSU rRNA gene. Using these regions, isolates of the *Alexandrium tamarense/fundyense/ catenella* species complex were shown to be related by geographic origin rather than by morphological affinities (Scholin et al., 1994b), which was originally indicated by the isozyme analysis. The world-wide biogeographic dispersal of ancestral population from the Pacific into the Atlantic has been hypothesized from these data. Furthermore, *Alexandrium* isolates will interbreed more successfully if they have similar isozyme patterns from two different locations than will isolates from the same locations but with different isozyme patterns (Sako et al., 1990). We now suspect that in these areas where isolates do not interbreed, they likely originate from different geographic clades that are overlapping in their distribution. For example, on the east coast of the UK down to about the Firth of Forth along the North Sea coast of Scotland, the non-toxic Western European clade of *Alexandrium tamarense* will overlap with the toxic North American clade. In contrast, other dinoflagellates, such as isolates of *Gambierdiscus toxicus* from similar geographical regions were not shown to be closely related, which suggested a multiclonal origin (Chinain et al., 1997). Populations of the green freshwater alga, *Gonium pectorale,* also appear from several locations to be multiclonal (Sako et al., 1991).

Microsatellites (MS) or simple sequence repeats (SSR), are the most powerful molecular markers available (Burke et al., 1991; Wright et al., 1994). In the beginning, MS were mainly from the field of fisheries sciences with most if not all economically important fish and shellfish species covered, but by now microsatellite markers and their use are available for macroalgae (e.g., *Gracilaria gracilis, Laminaria digitata*) and microalgae (e.g., *Chlamydomonas reinhardtii, Emiliania huxleyi, Ditylum brightwellii,* various *Pseudo-nitzschia* and *Alexandrium* species). It is from the microalgal studies that we find the strongest evidence for fragmentation of oceanic populations.

Microsatellites are short sequences of one to six nucleotides, e.g., $(CT)_n$ or $(CAG)_n$, that are repeated five to dozens and sometimes hundreds of times and are found in great abundance dispersed all over the genomes of all organisms investigated so far. This abundance together with the large number of alleles, resulting from high mutation rates because of their special, regular structure, makes them highly useful molecular markers at the population level. Microsatellite polymorphisms can be revealed where other marker types have failed and therefore they are especially useful for species that otherwise lack a high degree of polymorphism, such as inbreeding species like important crops as soybean, or clonal species like planktonic algae that do not have a regular sexual cycle. Comparisons of different marker types have shown that microsatellites have the highest degree of polymorphism of all commonly used marker types. Both genetic diversity and gene flow can be calculated from this marker. Microsatellite markers usually fail to resolve any genetic structure only when the populations are very recently diverged and in this case AFLPs will usually provide better resolution (Alpermann 2009).

All MS studies have shown high genetic diversity in planktonic populations (see review in Medlin et al. 2000). Rynearson and Armbrust (2000, 2004, 2005, 2006) studied the diatom *Ditylum brightwellii* in the Puget Sound estuary. Four genetically distinct and highly diverse populations were identified that differed in the timing and localisation within the estuary over the course of seven years. Distinct physiological characteristics were associated with each genetically distinct population. Genetically distinct populations in the upper basin of the estuary were never found in the lower basin of the estuary despite a constant flushing rate from the upper basin to the lower basin. In a study of more localised area, the flagellate *Heterosigma akashiwo* around Japan was composed of distinct populations with little evidence for gene flow between them even though tidal currents would permit natural dispersal of the cells from one area to another (Nagai et al., 2007). The global cosmopolitan coccolithophore, *Emiliania huxleyi* is highly diverse (Iglesias-Rodriguez et al., 2006) with disjunct global populations and little gene flow between populations in North Atlantic and Norwegian fjords. The Norwegian fjords were resampled 10 years apart with a shift in the genetic structure and only one genotype being shared by the population sampled in 1990 and in 2000. An estimate of the number of unique genotypes of *E. huxleyi* on a global basis was 9.4×10^{20}, a number scarcely believable when most oceanographers think that blooms are clonal and modellers only use one strain of a species in their models for climate change. More recently, the level genetic polymorphism of one phytoplanktonic eukaryote, *Ostreococcus tauri* has been estimated in the Gulf of Lion by using a population genomic approach to target neutral evolving genomic regions (Piganeau et al., 2010) that showed no spatial structure of these species in the Gulf of Lion and provided evidence for recombination in the ancestry of 17 isolates. With the development of Next Generation Sequencing, genetic diversity of whole communities will be available from metagenomic data. The planktonic cosmopolitan diatom *Pseudo-nitzschia multiseries* also contains genetically distinct and highly diverse and distinct gene pools between North American and European populations (Evans et al., 2004), whereas a morphologically similar cosmopolitan species, *Pseudo-nitzschia pungens*, is also highly diverse but with little population structure in local areas (Evans et al., 2005) but globally with distinct populations corresponding to major oceanic water masses (Casteleyn et al., 2010). However, all global isolates can interbreed (Chepurnov et al., 2005) and thus this species is the only example of a planktonic protist so far tested with a global gene pool with distinct population structure. In the toxic dinoflagellate, *Alexandrium tamarense*, microsatellites revealed four populations in the study area around the Orkney Islands, which were assumed to be temporal populations that had resulted form the inoculation of different year classes from the cyst beds in the current year's bloom (Alpermann 2009). All clones were phenotypically distinct. In the freshwater diatom, *Sellaphora capitata*, MS revealed that only a small number of alleles from water bodies in Scotland, England, Belgium and Australia could be found in all isolates (Evans et al., 2009), indicating a limited dispersal between populations, although all isolates could still interbreed. In the antarctic haptophyte *Phaeocytis antarctica*, each gyre in around the Antarctic had a unique genonotype and the Antarctic Circumpolar Current disperses genotypes from one gyre to another (Gaebler-Schwarz 2009). It is obvious that not only lakes but also oceans have genetically distinct populations with varying amounts of gene flow between them, some separated temporally and others spatially (Medlin 2007).

5. Molecular probes for identification and characterization of marine phytoplankton

Quite often morphological features as seen by light microscopy are not sufficient to distinguish clearly between species or groups of phytoplankton or marine bacteria. Therefore, more expensive methods, such as electron-microscopy or analysis of specific chemical components by HPLC, are needed to identify with certainly any species, but these are laborious, time-consuming and expensive. An alternative approach is to use specific molecular probes. Probes are short oligonucleotides of normally 16-24 bp length that are hundred percent homologous only to a complementary sequence in a gene of the species of interest and differ by at least one position to all other organisms. In hybridisation experiments, these probes can therefore be used to identify species of interest by binding to the target's sequence and later detection by a probe-attached label, e.g., Digoxigenin (DIG) or a fluorochrome like Fluorescein (FITC, Fig. 1). The application range of these probes extends from answering ecological questions, such as species composition and its change through space and time to the development of an early warning system for harmful algal blooms using probes for toxic species.

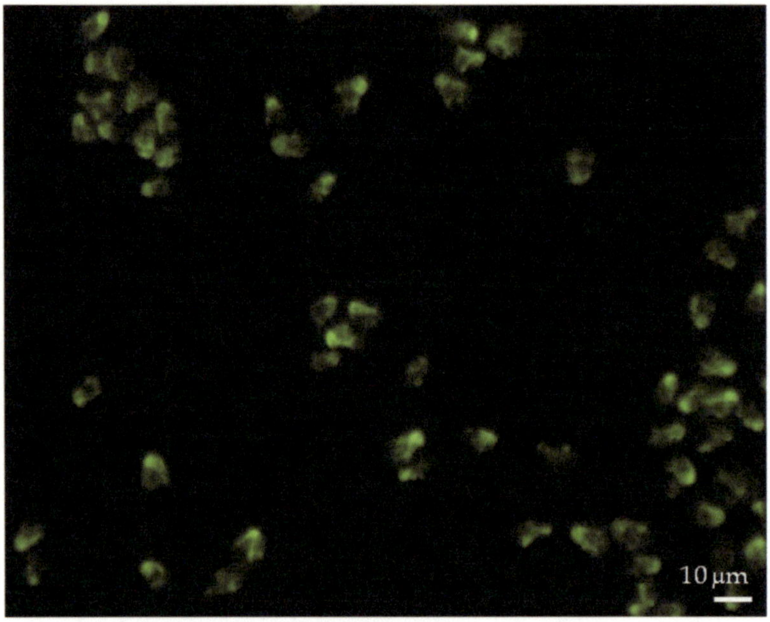

Fig. 1. FISH of the pear shaped toxigenic dinoflagellate *Azadinium spinosum*, x40.

The use of rDNA sequences has also other advantages for probe design. First, this molecule has regions with different degrees of conservation, which makes it possible to develop probes for higher taxonomic groups (class level probes, e.g., for prymnesiophytes Simon et al., 2000, probes for groups of related species ("clades") [clades of toxic or non-toxic *Chrysochromulina/Prymnesium* species (Simon et al., 2000), genus-specific probes for *Phaeocystis* species (Lange et al., 1996) down to species- or even strain-level probes for *Chrysochromulina polylepis* (Simon et al., 2000) and the toxic North American clade of *Alexandrium tamarense* (John et al., 2003). This hierarchical approach makes it easier to analyze field samples because higher level probes can be applied to the samples and then, depending on these results, only probes of a corresponding lower level need be used, therefore, reducing the number of necessary experiments. Because of the limited number of fluorochormes, usually two different, e.g., FITC and Cy5, are all that can be used in a single experiment especially if the taxa under investigation are photosynthethic. Second, the use of probes for rDNA allows them also to bind to the rRNA of ribosomes *in situ*, making it possible to use fluorochrome-labelled probes in whole-cell hybridisation experiments (FISH). The thousands of ribosomes provide enough targets for probe binding and therefore, strong enough signals to be detected. If this is not the case, i.e., in picoplankton and also in bacterial cells, which often show weaker signals because of their small size and therefore lower ribosome content, techniques like catalyzed reporter deposition-fluorescence *in situ* hybridization (CARD-FISH, Fig. 2) can be used to boost the signal strength up to a detectable level (Schönhuber et al., 1997; 1999). This method combined with FISH increases the intensity of fluorescence and thus raises the detection limit and the signal/noise ratio, which is critical for small cells and results in a strong signal enhancement of the hybridized cells up to 20 times compared to probes with a single fluorochrome (Fig. 2). CARD-FISH has been shown to be very useful in the detection of cyanobacteria (Schönhuber et al., 1997; 1999; West et al., 2001), picoplankton cells (Biegala et al., 2003, Not et al., 2004; 2002) and bacteria associated with micro algae (Biegala et al., 2002, Alverca et al., 2002).

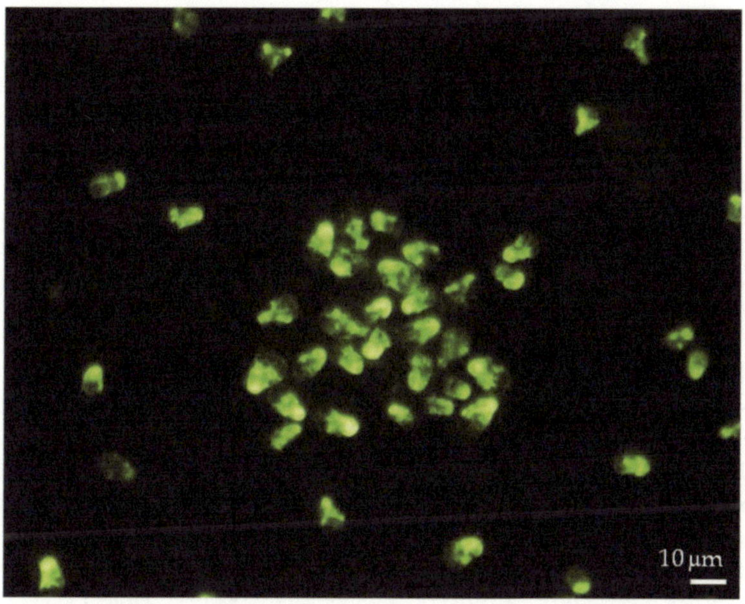

Fig. 2. CARD-FISH of the pear shaped toxigenic dinoflagellate *Azadinium spinosum*, x40.

Fluorescence *in situ* hybridisation targeting ribosomal RNA molecules has been often used for identification of harmful species in field samples. This molecular biological detecting tool is often deployed together with other molecular biological techniques, like quantitative PCR, to visualize the morphology of harmful algal species and for morphological comparisons using traditional methods, such as Utermöhl counts. Another important benefit of (CARD-) FISH is the potential detection and morphological visualization of unculturable harmful microalgae, such as the diarrhoeic shellfish poising causing dinoflagellate *Dinophysis*.

Higher group level probes, species, or even strain specific oligonucleotide probes for FISH are available for many taxa and enable the differentiation of morphological similar co-occurring species, especially for harmful algal bloom (HAB) species. Probes are available for detecting toxic species of the marine pennate diatom genus *Pseudo-nitzschia* that are associated with domoic acid in natural samples (e.g., Scholin et al., 1996 & 1997) as well as for toxic species in the important bloom forming microalgae, *Chrysochromulina* (Simon et al., 2000) and *Prymnesium* (Simon et al., 2000; Töbe et al., 2006) from marine and brackish environments. Probes targeting the different species of the toxigenic paralytic shellfish toxin (PSP) producing dinoflagellate genus *Alexandrium* include those detecting the North American *Alexandrium tamarense* ribotype (Miller and Scholin 1998; John et al., 2003), the toxic Temperate Asian ribotype (Hosoi-Tanabe and Sako 2005), the non-toxic Western European ribotype (John et al., 2005, Touzet et al. 2008), the non-toxic Mediterranean species complex (John et al., 2005), *A. minutum*, the non-toxic co-occurring morphologically similar species *A. andersoni* in Irish coastal waters (Touzet and Raine 2007; Touzet et al. 2008) and *A. peruvianum* (Touzet et al., 2011) and the spirolide producing taxa *A. ostenfeldii* (John et al., 2003, Touzet et al., 2011) All of these probes have been applied in field studies (John et al., 2005; Anderson et al., 2005). A set of FISH probes are also developed (John et al., in prep.) for the newly described genus *Azadinium*, (Tillmann et al., 2009) which comprises three different species, the toxigenic species *A. spinosum* and the non-toxic representatives. *A. obesum* & the very recently described species *Azadinium poporum* (Tillmann et al. 2011). *Azadinium spinosum* has been shown as the culprit for Azaspiracids, the most recently discovered group of lipophilic marine biotoxins of microalgal origin associated with human incidents of shellfish poisoning (Tillman et al., 2009 & 2011).

Fixation is a critical point in FISH applications and a suitable fixation method for species that are difficult to fix should be developed prior to any FISH application The harmful radiophyte *Heterosigma akashiwo*, which causes mass mortalities of cultured fish, alters its morphology dramatically after fixation with various commonly used fixatives. Chen et al. (2008) found that the gentle, but effective saline ethanol fixation method described by Miller and Scholin (1996 & 2000) is suitable for the fragile cells of *H. akashiwo* and does not cause clumping or breaking of the cells as does formalin or glutaraldehyde fixation. However, even this gentle fixation method can slightly change the cell's morphology, but this distorted morphology does not have a negative effect on the FISH signals of the newly developed species specific probes for *H. akashiwo* (Chen et al., 2008).

CARD-FISH has been deployed to a lesser extent in HAB studies, most probably because of the higher cost of the labelled probe, the chemicals used for signal enhancement and the additional needed time for the signal enhancement. Rehnstam-Holm et al. (2002) developed a genus specific probe for *Dinophysis* and applied it successfully in CARD-FISH experiments with field samples containing different *Dinophysis* species. This FISH technique has considerably alleviated a persistent autofluorescence of species, such as that exhibited by

Prymnesium parvum cells (Töbe et al., 2006) or to detect ingested cells in the guts of other species.

Despite the numerous advantages of FISH applications in HAB studies, the use of other techniques, such as high sample throughput techniques to analyse bulk environmental samples in a shorter time have not been widely applied to field samples. One such method of detection is by flow cytometry, which is not *per se* a molecular biological method, but can be used in combination with molecular probes to great advantage to analyze large numbers of cells (Wallner et al., 1993). CARD-FISH has been successfully applied for the identification and enumeration of phytoplankton cells by flow cytometry (Biegela et al., 2003). Probes have been used with the solid phase cytometer to scan and enumerate all cells on a filter (*Prymnesium parvum* in Töbe et al., 2006, cryptomonads in Medlin & Schmidt 2010). In Medlin & Schmidt (2010), a hierarchical probe approach was used to study the cryptomonads in Arcachon Bay, France and they were able to show that genera belonging to Clade 3 were dominant in this bay system.

When extracted DNA is available, another method for the use of oligonucleotide probes is as PCR primers. A specific oligonucleotide in combination with a matching primer from a highly conserved region of the same gene should only amplify a product if the DNA comes from the species for which the oligonucleotide probe was designed. Nevertheless, when a probe can be used this way, this method is much faster than a dot blot hybridisation in detecting the presence of a certain type of organism, which can even be quantified through the use of quantitative real-time PCR (qPCR).

qPCR enables a high sample throughput and several species can be detected at a time even when only small sample volumes are available. Data are collected over the entire PCR cycle by using fluorescent markers that are incorporated into the PCR amplicon during amplification and directly in the exponential phase where PCR is precise, thus avoiding the problem of the amplification plateau of qualitative PCR experiments. The increasing fluorescence is measured and the change in fluorescence is directly proportional to the amount of starting material (Demir et al. 2008). Different qPCR chemistries are available, depending on fluorescent dyes binding to double stranded DNA (dsDNA) or the application of fluorescently labelled species-specific oligonucleotide probes. SYBR Green is the most commonly used methods used in qPCR applications using primers specific for the target DNA. In another more sensitive and specific approach primers together with a specific fluorigenic oligonucleotide probe are used. This probe based qPCR approaches enables the detection of several different original templates in one sample, whereby the number of detectable target genes in one sample is limited by the number of available fluorescence reporter dyes for the separate probes, which can be excited by the qPCR instrument. However, these multiplex qPCR experiments have to be carefully optimized (Kudela et al. 2010). However, results of qPCR experiments could be hampered by external influences, e.g., different DNA extraction yields depending on the extraction method used and the presence of humic substances that could influence or even inhibit the PCR reaction, possibly resulting in discrepancies between traditional cell counts and qPCR determined cell counts. Therefore, the (quantitative) species composition of the investigated habitat could be incorrectly recorded. These problems could be resolved or at least minimized by applying a standardized DNA isolation method generating high quality DNA samples and the use of an internal standard in some of the environmental samples to monitor the amplification efficiency of the qPCR experiment.

There are already examples where the combination of real-time PCR and species specific primers/probes has been successfully applied, e.g., for the detection and enumeration of the

A. catenella/fundyense/tamarense species complex (Dyhrman et al. 2006 & 2010), and with the rapid progress in technology, this will likely be a promising method for routine monitoring of selected species.

The previously described techniques are powerful and highly quantitative tools for the identification of microbial organisms. However, they all have the drawback that they are single probe approaches that are limited to the analysis of only one or a few targets at a time. The introduction of the concept of DNA microarrays about ten years ago suggests an option to void the limitations of single probe approaches. DNA microarray-experiments are multiplexed assays that provide the possibility for high throughput analysis of molecular probe based species identification without a cultivation step. This is of special interest for the identification of prokaryotic and eukaryotic cells with very small sizes and few distinct morphological features. This provides an aid to science because taxonomists are not being trained because this field is perceived to be no longer needed with the new age of molecular analyses. Therefore, DNA microarrays might be of special value for phycological studies because they represent a tool that does not require a broad taxonomic knowledge to identify cells. Consequently there are a growing number of publications that report the use of microarrays bearing molecular probes that target the rRNA for the identification of microbial species. They have been used successfully in combination with an amplification of the rRNA-gene for the identification of phytoplankton, bacteria, bacterial fish pathogens, and sulfate reducing prokaryotes (Gescher et al.,, 2008, Manz et al., 1992, Giovanonni et al., 1990, Pace et al., 1986, Rehnstam et al., 1993). DNA-microarrays allow the parallel analysis of almost infinite numbers of probes at a time in just one experiment. The technology is based on a DNA-microchip that contains an ordered array of molecular probes on its surface (Fig. 3).

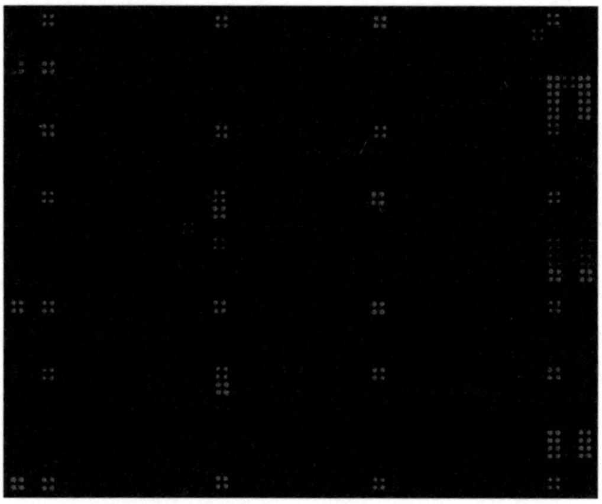

Fig. 3. Image of a scanned DNA-microarray from field sample taken in OsloFjord, Norway. Each cluster of 4 dots represents replicate probes specific for one species spotted onto the glass slides and hybridised to fluorescently labelled RNA from the field sample. The DNA-chip contained probes for various toxic phytoplankton taxa. Photo taken by Dr. S. Dittami for the EU MIDTAL project.

All these methods have, despite their high set up costs for the machines, the disadvantage of requiring quite bulky pieces of equipment, hence, making them difficult to be used in the field and on-board ship. This is of particular interest for the monitoring programs for toxic algae, where samples are taken regularly at places that are not in close proximity of laboratories that host the previously described equipment. DNA-biosensors could serve the needs of monitoring programs for an easy-to-handle and inexpensive tool. Electrochemical readings of DNA-biosensors are unambiguous and even for a scientific layperson easy to use and interpret. There are a number of examples for the application of DNA-biosensors that have been developed for the identification of organisms, including toxic algae. The application of a DNA-biosensor for the identification of organisms is again based on taxon specific molecular probes, of which a large number already exists for toxic algae.

A DNA-biosensor has been adapted for the electrochemical detection of the toxic dinoflagellate *Alexandrium tamarense, ostenfeldii* and *minutum* (Diercks et al., 2008, Metfies et al., 2005; Diercks et al. & 2011, respectively). The DNA-biosensor detection reaction is a sandwich-hybridisation that takes place on a carbon electrode on a disposable chip. A sandwich-hybridisation is based on a set of two specific molecular probes that bind in close proximity to the target nucleic acid. One probe, termed the capture probe, is immobilised via biotin on the carbon electrode, which is coated with avidin. The second probe, termed the signal probe, mediates the detection reaction, if target DNA is captured by hybridisation to the capture probe on the carbon electrode. The second probe is recognized by an antibody that is coupled to a horseradish-peroxidase that catalyses the red/ox reaction of hydrogen peroxide to water. The electron-transfer during the red/ox-reaction can be measured as a current, which is only possible if the target nucleic acid as a link between the capture and signal probes is present in the system. Experiments with RNA isolated from laboratory strains showed, that the electrochemical signal is proportional to the amount of target RNA applied to the sensor. The device was expanded in the EU ALGADEC project to regional chips for up to 14 harmful algal species at a time. In order to serve the needs of the monitoring programs that aim to count all potentially harmful algae in a geographic area it would be indispensable to adapt the present DNA-biosensor to a bigger range of toxic algae. This device is not yet available to the general public, nevertheless the method has a high potential to become a powerful monitoring tool in the future.

As it can be realised, there are numerous techniques possible for analysing multiple samples with specific probes automatically and more are surely to come. With them the way is open for mass screening of water samples for the detection of interesting marine species like toxic algae, even as there are still some problems to be solved and methods to optimise before they can be routinely used for this kind of purpose.

6. References

Alpermann, T. (2009). Evolutionary ecology of *Alexandrium* (Dinophyceae) with special emphasis on genotypic and phenotypic variation in the toxigenic species *A. tamarense*. PhD Thesis, University of Bremen

Alverca, E., Biegala, I.C., Kennaway, G.M., Lewis, J. & Franca, S. (2002). *In situ* identification and localization of bacteria associated with *Gyrodinium instriatum* (Gymnodiniales, Dinophyceae) by electron and confocal microscopy. *European Journal of Phycology, 37*, pp. 523–530

Amann, R.I. (1995). *In situ* identification of micro-organisms by whole cell hybridization with rRNA-targeted nucleic acid probes. In *Molecular Microbiology and Ecology Manual 3.3.6,* Akkermans, A.D.L., van Elsas, J.D., de Brujin, F.J., Eds., Kluwer Academic Publishers: Dordrecht, the Netherlands, 1995, pp. 1-15.

Andersen, R.A., Saunders, G.W., Paskind, M.P. & Sexton, J.P. (1993). Ultrastructure and 18S rRNA gene sequence for *Pelagomonas calceolata* gen. et sp. nov. and the description of a new algal class, the Pelagophyceae classis nov. *Journal of Phycology, 29,* pp. 701-715

Anderson, D.M., Kulis, D., Keafer, B.A., Gribble, K.E., Marin, R., II, & Scholin, C.A., 2005. Identification and enumeration of *Alexandrium* spp. from the Gulf of Main using molecular probes. *Deep-Sea Reseaerch* II, 52, pp. 2467-2490

Biegala, I.C., Kennaway, G., Alverca, E., Lennon, J.F., Vaulot, D. & Simon, N. (2002). Identification of bacteria associated with dinoflagellates (Dinophyceae), *Alexandrium* spp. using tyramide signal amplification-fluorescent *in situ* hybridization and confocal microscopy. *Journal of Phycology, 38,* pp. 404-411

Biegala, I.C., Not, F., Vaulot, D. & Simon, N. (2003). Quantitative assessment of picoeukaryotes in the natural environment by using taxon-specific oligonucleotide probes in association with Tyramide Signal Amplification-Fluorescence *in situ* hybridization and flow cytometry. *Applied and Environmental Microbiology,* 69, pp. 5519-5529

Brand, L.E. (1982). Genetic variability and spatial patterns of genetic differentiation in the reproductive rates of the marine coccolithophores *Emiliania huxleyi* and *Gephyrocapsa oceanica. Limnology and Oceanography, 27,* pp. 236-245

Brand, L.E. (1989). Review of genetic variation in marine phytoplankton species and the ecological implications. *Biological Oceanography, 6,* pp. 397-409

Brand, L.E., Murphy, L.S., Guillard, R.R.L. & Lee, H.-T. (1991). Genetic variability and differentiation in the temperature niche component of the diatom *Thalassiosira pseudonana. Marine Biology, 62,* pp. 103-110

Brazelton, W.J., Ludwig, K.A., Sogin M.L., Ekaterina N., Andreishcheva, N., Kelley, D. S., Chuan-Chou Shen, C-C., Edwards, R.L. & Barossa, J.A. (2010). Archaea and bacteria with surprising microdiversity show shifts in dominance over 1,000-year time scales in hydrothermal chimneys. *Proceedings of the National Academy of Sciences of the United States of America.* /doi/10.1073/pnas.0905369107

Bull, A. (2004). Microbioal diversity and bioprospecting. ASM Press, Waskington D.C. 497 pp

Burke, T., Hanotte, O. & Bruford, M.W. (1991). Multilocus and single locus minisatellite analysis in population biological studies. In: *DNA Fingerprinting, Approaches and Applications,* Burke, T., Dolf, G., Jeffreys, A.J. & Wolff, R., (Eds.), 154-68. Birkhaeuser: Basel, Switzerland

Casteleyn, G., Leliaert, F., Backeljau, T., Debeer, A.E., Kotaki, Y., Rhodes, L., Lundholm, N., Sabbe, K., Vyverman, W. (2010). Limits to gene flow in a cosmopolitan marine planktonic diatom. Proceedings of the National Academy of Sciences of the United States of America, doi/10.1073/pnas.1001380107.

Cembella, A.D. & Taylor, F. J. R. (1986). Electrophoretic variability within the *Protogonyaulax tamarensis/catenella* species complex, Pyridine linked dehydrogenases. *Biochemical Systematics and Ecology,* 143, pp. 311-323

Cembella, A.D., Taylor, F.J.R. & Therriault, J.C. (1988). Cladistic analysis of electrophoretic variants within the toxic dinoflagellate genus *Protogonyaulax. Botanica Marina, 31,* pp. 39-51

Chen, G.F., Wang, G.C., Zhang, C.Y., Wang, X.K. & Zhou, B.C. (2008). Development of rRNA and rDNA-targeted probes for the fluorescence in situ hybridization to detect *Heterosigma akashiwo* (Raphidophyceae). *Journal of Experimental Marine Biology and Ecology*, 355, pp. 66-75

Chepurnov, V.A., Mann, D.G., Sabbe, K., Vannerum, K., Casteleyn, G., Verleyen, E., Peperzak, L., Vyverman, W., 2005. Sexual reproduction, mating system, chloroplast dynamics and abrupt cell size reduction in *Pseudo-nitzschia pungens* from the NorthSea (Bacillariophyta). *European Journal of Phycology* 40, pp. 379–395.

Chinain, M., Germain, M., Sako, Y., Pauillac, S. & Legrand, A.-M. (1997). Intraspecific variation in the dinoflagellate *Gambierdiscus Toxicus* Dinophyceae, I. Isozyme analysis. *Journal of Phycology, 33*, 36-43

Chisholm, S.W., Olson, R.J., Zettler, E.R., Goericke, R., Waterbury, J. & Welschmeyer, N. (1988). A novel free-living prochlorophyte abundant in the oceanic euphotic zone. *Nature, 334*, pp. 340-343

Cuvelier, M.L., Allen, A.E., Monier, A., McCrow, J.P., Messie, M., Tringe, S.G., Woyke, T., Welsh, R.M., Ishoey, T., Lee, J.H., Binder, B.J., DuPont, C.L., Latasa, M., Guigand, C., Buck, K.R., Hilton, J., Thiagarajan, M., Caler, E., Read, B., Lasken, R.S., Chavez, F.P., Worden, A.Z. (2010). Targeted metagenomics and ecology of globally important uncultured eukaryotic phytoplankton. *Proceedings of the National Academy of Sciences, USA* 107, pp. 14679–14684.

Demir, E., Coyne, K.J., Doblin, M.A., Handy, S.M. & Hutchins, D.A. (2008). Assessment of Microzooplankton Grazing on *Heterosigma akashiwo* Using a Species-Specific Approach Combining Quantitative Real-Time PCR (QPCR) and Dilution Methods. *Microbial Ecology*, 55, pp. 583-594

Diercks, Sonja, Metfies, Katja, Jäckel, Steffi and Medlin, Linda K. 2008. Development and adaptation of a multiprobe biosensor for the use in a semi-automated device for the detection of toxic algae, Biosensors and Bioelectronics, 23: 1527-1533.

Diercks, S., Metfies, K., Jäckel, S. and Medlin, L.K. 2011. Development and optimisation of a semi-automated rRNA biosensor for the detection of toxic algae. Harmful Algae, in press.

Dyhrman, S.T., Erdner, D., La Du, J., Galac, M. & Anderson, D.M. (2006). Molecular quantification of toxic *Alexandrium fundyense* in the Gulf of Maine using real-time PCR. *Harmful Algae*, 5, pp. 242-250

Dyhrman, S.T., Haley, S.T., Borchert, J.A., Lona, B., Kollars, N. & Erdner, D.L. (2010). Parallel Analyses of *Alexandrium catenella* Cell Concentrations and Shellfish Toxicity in the Puget Sound. *Applied and Environmental Microbiology*, 76, pp. 4647-4654

Evans, K.M., Bates, S.S., Medlin, L.K. & Hayes P.K. (2004). Microsatellite marker development and genetic variation in the toxic marine diatom *Pseudo-nitzschia multiseries* (Bacillariophyceae), *Journal of Phycology, 40*, pp. 911-920

Evans, K.M., Kühn, S.F. & Hayes, P.K. (2005). High levels of genetic diversity and low levels of genetic differentiation in North Sea *Pseudo-nitzschia pungens* (Bacillariophyceae) populations, *Journal of Phycology*, 41, pp. 506-514

Evans, K.M., Wortley, A.H.& Mann, D.G. (2007). An assessment of potential Diatom "Barcode" Genes: *coxI*, *rbc*L, 18S and ITS rDNA, and their effectiveness in determining relationships in *Sellaphora* (Bacillariophyta). *Protist, 158*, pp. 349-364

Evans, K.M., Chepurnov, V.A., Sluiman, H.J., Thomas, .S.J., Spears, B.M. & Mann, D.G. (2009). Highly differentiated populations of the freshwater diatom *Sellaphora capitata* suggest limited dispersal and opportunities for allopatric speciation. *Protist, 160*, pp. 386–396

Gaebler-Schwarz, S. (2009). Estimation of genetic diversity in the colony forming polar pymnesiophyte species *Phaeocystis antarctica*. PhD. Thesis, University of Bremen, pp. 2217.

Gaebler, S., Hayes, P.K. & Medlin, L.K. (2007). Methods used to reveal genetic diversity in the colony forming prymnesiophyte *Phaeocystis antarctica* - preliminary results. *Biogeochemistry*, 83, pp. 19-27

Gallagher, J.C. (1980). Population genetics of *Skeletonema costatum* (Bacillariophyceae), in Narragansett Bay. *Journal of Phycology*, 16, pp. 464-74

Gescher, G., Metfies K., Frickenhaus, S., Knelfelkamp, B., Wiltshire, K. and Medlin, L.K (2008). Feasibility of assessing community composition of Prasinophytes at the Helgoland Reede Sampling Site with a DNA-Microarray. *Applied and Environmental Microbiology*, 74, pp. 5305-5316.

Giovanonni, S.J., Britschgi, T.B., Moyer, C.L.& Field, K.G. (1990). Genetic diversity in Sargasso Sea bacterioplankton. *Nature*, 345, pp. 60-63

Gosling, E.M. Speciation and species concepts in the marine environment, In *Genetics and Evolution of Aquatic Organisms*, Beaumont, A.R., Ed., Chapman & Hall: London, U.K. 1994, pp.1-15.

Grassle, G.F., Lasserre, P. McINtyre, A.D., & Ray, G.C. (1991). Marine Biodiversity and Ecosystem Function. *Biology International* Special Issue 23, pp 1-18

Guillou, L., Chretiennot-Dinot, M-J., Medlin, L.K., Claustre, H., Loiseaux-De Goer, S. & Vaulot, D. (1999). *Bolidomonas*, a new genus with two species belonging to a new algal class, The Bolidophyceae. *Journal of Phycology, 35*, pp. 368-381

Hadrys, H., Balick, M., Schierwater, B. Application of random amplified polymorphic DNA RAPD, *Mol. Ecol.* 1992, *1*, 55-63

Hasle, G.R. (1978). Some fresh water and brackish water species of the diatom genus *Thalassiosira* Cleve. *Phycologia, 17*, pp. 263-292

Hasle, G.R. (1983). The marine planktonic diatom *Thalassiosira oceanica* sp. nov. and *T. parthenia*. *Journal of Phycology, 19*, pp. 220-229

Hayhome, B.A., Anderson, D.M., Kulis, D.M. & Whitten, D.J. (1989). Variation among congeneric dinoflagellates from the northeastern United States and Canada. I. Enzyme electrophoresis. *Marine Biology*, 101, pp. 427-435

Hayhome, B.A., Whitten D.J., Harkins, K.R. & Pfiester, L.A. (1987). Intraspecific variation in the dinoflagellate *Peridinium volzii*. *Journal of Phycology*, 23, pp. 573-580

Hedgecock, D. (1994). Population genetics of marine organisms. *US GLOBEC News, 6*, pp. 1-3, 11

Hosoi-Tanabe, S. & Sako, Y. (2005). Rapid detection of natural cells of *Alexandrium tamarense* and *A. catenella* (Dinophyceae) by fluorescent in situ hybridization. *Harmful algae*, 4, pp. 319-328

Iglesias-Rodriguez, M.D., Schofield, O.M., Batley, P.J., Medlin, L.K. &Hayes P.K. (2006). Extensive intraspecific genetic diversity in the marine coccolithophore *Emiliania huxleyi*: the use of microsatellite analysis in marine phytoplankton population studies. *Journal of Phycology*, 42, pp. 526-536

John, U., Fensome, R.A. &Medlin, L.K. (2003). The application of a molecular clock based on molecular sequences and the fossil record to explain the biogeographic distribution within the *Alexandrium tamarense* "species complex". *Molecular Biology Evolution*, 20, pp. 1015-1027

John, U., Groben, R., Beszteri, B. & Medlin, L.K. (2004). Utility of Amplified Fragment Length Polymorphisms (AFLP) to analyse genetic structures within the *Alexandrium tamarense* species complex. *Protist*, 155, pp. 169-179

John, U., Medlin, L. K. & Groben, R. (2005). Development of specific rRNA probes to distinguish between geographic clades of the *Alexandrium tamarense* species complex. *Journal of Plankton Research*, 27, pp. 199-204

Killan, B.P. & Gaines, S.D. (2003). Propagule dispersal in marine and terrestrial environments: a community perspective*ecology*, 84 pp. 2007-2020

Kim, K., Harrison, J.W., Sudek, S., Jones, M.D.H., Wilcox, H.M., Richards, T.A., Worden, A.Z. & Archibald, J.M. (2011). Newly identified and diverse plastid-bearing branch on the eukaryotic tree of life. *Proceedings of the National Academy of Sciences, USA* 108, doi/10.1073/pnas.1013337108

Kudela, R.M., Howard, M.D.A., Jenkins, B.D., Miller, P.E. and Smith, G.E. (2010). Using the molecula toolbox to compare harmful algal blooms in upwelling systems. *Progress in Oceanography*, 85, pp. 108-121

Lami, R., Ghiglione, J-F., Desdevieses, Y., West, N., Lebaron, P. (2009). Annual patterns of presence and activity of marine bacteria monitored by 16S rDNA-16SrRNA fingerprints in teh coastal NW Mediterranean Sea. *Aquatic Microbial Ecology, 54*, pp. 199-210

Lange, M., Simon, N., Guillou, L., Vaulot, D., Amann, R., Ludwig, W. & Medlin, L.K. (1996). Identification of the Class Prymnesiophyceae and the genus *Phaeocystis* with rRNA-targeted nucleic acid probes detected by flow cytometry. *Journal of Phycology*, 32, pp. 858-868

Lynch, M. & Milligan, G. (1994). Analysis of population genetic structure with RAPD markers. *Molecular Eco*logy *3*, pp. 91-99

Manz, W., Amann, R., Ludwig, W., Wagner, M., Schliefer, K-H. (1992). Phylogenetic oligodeoxynucleotide probes for the major subclasses of proteobacteria, problems and solutions. *Systematics and Applied Microbiolbiology* 15, pp. 593-600.

Medlin, L. K. 2007. If everything is everywhere, do they share a common gene pool? *Gene*, 405, pp 180-183

Medlin, L.K. and Kooistra, W.H.C.F. (2010). A review of the diversity in the marine photosynthetic protist community and the methods used to estimate this biodiversity. Special Issue of Diversity "Biological Diversity Assessed by Molecular Methods", *2*, pp. 973-1014.

Medlin, L.K., Lange, M. & Noethig, E.V. (2000). Genetic diversity of marine phytoplankton, A review and a look to Antarctic phytoplankton. *Antarctic Science, 12*, pp. 325-331

Medlin, L.K & Schmidt, K. (2010). Molecular probes improve the taxonomic resolution of cryptophyte abundance in Archachon Bay. *Vie et Millieu* 60, pp 9-15

Metfies, K., Huljic, S., Lange, M. & Medlin, L.K. (2005). Electrochemical detection of the toxic dinoflagellate *A. ostenfeldii* with a DNA Biosensor. *Biosensors and Bioelectronics*, 20, pp. 1349-1357

Miller, P.E. & Scholin, C.A. (1996). Identification of cultured *Pseudo-nitzschia* Bacillariophyceae, using species-specific LSU targeted fluorescent probes. *Journal of Phycology, 32*, pp. 646-655

Miller, P.E. & Scholin, C.A. (1998). Identification and enumeration of cultured and wild *Pseudo-nitzschia* (Bacillariophyceae) using species specific LSU rRNA –targeted fluorescent probes and filter-based whole cell hybridization. *Journal of Phycology*, 34, pp. 371-382

Miller, P. E. & Scholin, C. A. (2000) On detection of *Pseudo-nitzschia* (Bacillariophyceae) species using whole-cell hybridization: Sample fixation and stability. *Journal of Phycology*, 36, pp. 238-250

Moestrup, Ø. (1991). Further studies of presumedly primitive green algae, including the description of Pedinophyceae class. nov. and *Resultor* gen. nov. *Journal of Phycology*, 27, pp. 119-133

Murphy, L.S. & Guillard, R.R.L. (1976). Biochemical taxonomy of marine phytoplankton by enzyme electrophoresis. I. The centric diatoms *Thalassiosira pseudonana* Hasle Heimdal and *Thalassiosira fluviatilis* Hustedt *Journal of Phycology*, 12, pp. 9-13

Muyzer, G., De Wall, E.C., & Uitterlinden, A.G. (1993). Profiling on complex microbial populations by denaturing gradient gel electrophoresis analysis by polymerase chain reaction-amplified genes coding for 16S rRNA. *Applied and Environmental Microbiology, 59*, pp. 695-700

Nagai, S., Lian,C., Yamaguchi, S., Hamaguchi, M., Matsuyama, Y., Itakura, S., Shimada, H., Kaga, S., Yamauchi, H., Sonda, Y., Nishikawa, T., Kim, C.H., Hogetsu, T. (2007). Microsatellite markers reveal population genetic structure of the toxic dinoflagellates (Dinophyceae) in Japanese coastal waters. *Journal of Phycology*, pp. *43*, 43-54

Not, F., Valentin, K., Romari, K., Massana, R., Vaulot, D., Medlin, L.K. (2007). Picobiliphytes: A marine picoplanktonic algal group with unknown affinities to other eukaryotes. *Science, 315*, pp. 253-255

Ormond, R.F., Gage, J.D. & Angel, M.V. (1998). *Marine Biodiversity, Patterns and Processes*. Cambridge University Press, Cambridge, U.K., 449 pp.

Parker, P.G., Snow, A.A., Schug, M.D., Booton, G.C., Fuerst, P.A. (1998). What molecules can tell us about populations: choosing and using a molecular marker. *Ecology 79*, pp. 361-382

Pace, N.R., Stahl, D.A., Lane, D.J. & Olsen, G.J. (1986). The analysis of microbial populations with ribosomal RNA sequences. *Advances in Microbial Ecology*, 9, pp. 1-55

Palumbi, S. R. (1992). Marine speciation on a small planet. *TREE*, 7, pp. 114-118

Piganeau, G., Eyre-Walker, A., Grimsley, N. & Moreau, H. (2011) How and Why DNA Barcodes Underestimate the Diversity of Microbial Eukaryotes PLOS one, 6, e16342. doi:10.1371

Rehnstam, A.S., Bäckman, S., Smith, D.C., Azam, F. & Hagström, Å. (1993). Blooms of sequence-specific culturable bacteria in the sea. *Federation of European Microbiological Societies Microbiology Ecology*, 102, pp. 161-166

Rehnstam-Holm, A.S., Godhe, A. & Anderson, D.M. (2002). Molecular studies of *Dinophysis* (Dinophyceae) species from Sweden and North America. Phycologica, 41, pp. 348-357

Rynearson, T.A. & Armbrust, E.V. (2000). DNA fingerprinting reveals extensive genetic diversity in a field population of the centric diatom *Ditylum brightwellii. Limnology and Oceanography,* 45, pp. 1329-1340

Rynearson, T. A. & Armbrust, E. V. (2004). Genetic differentiation among populations of the planktonic marine diatom *Ditylum brightwellii* (Bacillariophyceae*), Journal of Phycology,* 40, pp. 34-43

Rynearson, T. A. & Armbrust, E. V. (2005). Maintenance of clonal diversity during a spring bloom of the centric diatom *Ditylum brightwellii. Molecular Ecology,* 14, pp. 1631-1640

Rynearson, T. A., Newton J.A. & Armbrust, E. V. (2006). Spring bloom development, genetic variation, and population succession in the planktonic diatom *Ditylum brightwellii. Limnology and Oceanography,* 51, pp. 1249-1261

Sako, Y., Kim, C.H., Ninomiya, H., Adachi, M. & Ishida, Y. (1990). Isozyme and cross analysis of mating populations in the *Alexandrium catenella/tamarense* species complex. In: *Toxic Marine Phytoplankton,* Graneli, E., Sundstrom, B., Edler, L. & D.M. Anderson, (Eds.), 320-329, Elsevier: New York, NY, USA

Sako, Y., Shrestha, K., Uchida, A. & Ishida, Y. (1991). Isozyme analysis of mating populations of *Gonium pectorale* Chlorophyta, *Journal of Phycology,* 27, pp. 309-315

Scholin, C.A. & Anderson, D.M. (1994a). Identification of group- and strain-specific genetic markers for globally distributed *Alexandrium* (Dinophyceae). I. RFLP analysis of SSU rRNA genes. *Journal of Phycology,* 29, pp. 209-216.

Scholin, C.A., Hallegraeff, G.M. & Anderson, D.M (1994b). Identification of group- and strain-specific genetic markers for globally distributed *Alexandrium* (Dinophyceae). II. Sequence analysis of a fragment of the LSU rRNA gene. *Journal of Phycology,*30, pp. 999-1011

Scholin, C.A., Hallegraeff, G.M. & Anderson, D.M. (1995). Molecular evolution of the *Alexandrium tamarense* species complex (Dinophyceae), dispersal in the North American and West Pacific regions. *Phycologia,* 34, pp. 472-485

Scholin, C., Miller, P., Buck, K. & Chavez, F. (1997). Detetion and quantification of *Pseudo-nitzschia australis* in cultured and natural populations using LSU rRNA-targeted probes. *Limnology and Oceanography,* 42, pp. 1265-1272

Simon, N., Campbell, L., Ornolfsdottir, E., Groben, R., Guillou, L., Lange, M. & Medlin, L.K. (2000). Oligonucleotide probes for the identification of three algal groups by fluorescent whole-cell hybridization. *Journal of Eukaryotic Microbiology,* 47, 76-84

Simon, N., Brenner, J., Edvardsen, B. & Medlin, L.K. (1997). The identification of *Chrysochromulina* and *Prymnesium* species Haptophyta, Prymnesiophyceae, using fluorescent or chemiluminescent oligonucleotide probes, a means for improving studies on toxic algae. *European Journal of Phycology,* 32, pp. 393-401

Sogin, M.L., Morrison, H.G., Huber, J.A., Welch, D.M., Huse, S.M., Neal, P.R., Arrieta, J.M. & Herndl, G.J.. (2006). Microbial diversity in the deep sea and the underexplored "rare biosphere". *Proceedings of the National Academic Science USA,* 103, pp. 12115-12120

Tillmann, U., Elbrächter, M., Krock, B., John. U. & Cembella, A. (2009). *Azadinium spinosum* gen. et sp. nov. (Dinophyceae) identified as a primary producer of azaspiracid toxins. *European Journal of Phycology* ,44, pp. 63-79

Tillmann, U., Ellbrächter, M., John, U. & Krock, B. (2011). A new non-toxic species in the dinoflagellate genus *Azadinium*: *A. poporum* sp. nov. *European Journal of Phycology,* 46, pp. 74-87Töbe, K., Eller, G. & Medlin, L.K. (2006). Automated detection and enumeration for toxic algae by solid-phase cytometry and the introduction of a new probe for *Prymnesium parvum* (Haptophyta: *Prymnesiophyceae*). *Journal of Plankton Research*, 7, pp. 643-657

Touzet, N. & Raine, R. (2007). Discrimination of *Alexandrium andersoni* and *A. minutum* (Dinophyceae) using LSU rRNA-targeted oligonucleotide probes and fluorescent whole cell hybridization. *Phycologica*, 46, pp. 168-177

Touzet, N.,Franco, J.M. & Raine, R. (2008). PSP toxin analysis and discrimination of the naturally co-occurring *Alexandrium tamarense* and *A. minutum* (Dinophyceae) in Cork Harbour, Ireland. *Aquatic Microbiology Ecology*, 51, pp. 285-299

Touzet, N., Keady, E., Raine, R. & Maher, M. (2009). Evaluation of taxa-specific real time PCR, whole-cell FISH and morphotaxonomy analyses for the detection and quantification of the toxic microalgae *Alexandrium minutum* (Dinophyceae), Global Clade ribotype. *Federation of European Microbiological Societies Microbiology Ecology*, 67, pp. 329-41

Touzet, N., Davidson, K., Pete, R., Flanagan, K., McCoy, G.R., Amzil, Z., Maher, M. Chapelle, A. & Raine, R. (2010). Co-Occurrence of the West European (Gr.II) and North American (Gr.I) Ribotypes of *Alexandrium tamarense* (Dinophyceae) in Shetland, Scotland. *Protist*, 161, pp. 370-384

Touzet, N., Lacaze, J.P., Maher, m., Turrell, E. & Raine, R. (2011). Summer dynamics of *Alexandrium ostenfeldii* (Dinophyceae) and spirolide toxins in Cork Harbour, Ireland. *Marine and Ecology Progress Series*, 425, pp. 21-33

Vos, P., Hogers, R., Bleeker, M., Reijans, M., Van De Lee, T., Hornes, M., Frijters, A., Pot, J., Peleman, J., Kuiper, M. & Zabeau, M. (1995). AFLP, a new technique for DNA fingerprinting. *Nucleic Acids Research*,23, pp. 4407-4414

Wallner, G., Amann, R. & Beisker, W. (1993). Optimizing fluorescent in situ hybridization with rRNA targeted probes for flow cytometric identification of microorganisms. *Cytometry*, 14, pp. 136-143

Waterbury, J.B., Watson, S.W., Guillard, R.R.L. & Brand, L.E. (1979). Widespread occurrence of a unicellular, marine planktonic cyanobacterium. *Nature London*, 345, pp. 63-65

Waterbury, J.B., Watson, S.W., Valois, F.W.I. & Franks, D.G. (1986). Biological and ecological characterization of the marine unicellular cyanobacterium *Synechococcus*, in photosynthetic picoplankton. In: *Canadian Bulletin of Fisheries and Aquatic Sciences*, Platt, T., & W.K.W. Li, (Eds.), 214, pp. 71-120

West, N.J., Schönhuber, W.A., Fuller, N., Amann, R.I., Rippka, R., Post, A .& Scanlan, D. (2001). Closely related *Plochlorococcus* genotypes show remarkably different depth distributions in two oceanic regions as revealed by in situ hybridization using 16SrRNA-targeted oligonucleotides. *Microbiology*, 47, pp. 1731-1744

Wright, J.M. & Bentzen, P. (1994).Microsatellites, genetic markers for the future. *Reviews of Fish Biology and fisheries*, 4, pp. 384-388